Principles of Mobile Computing and Communications

Principles of Mobile Computing and Communications

Mason Bell

New York

Published by NY Research Press
118-35 Queens Blvd., Suite 400,
Forest Hills, NY 11375, USA
www.nyresearchpress.com

Principles of Mobile Computing and Communications
Mason Bell

International Standard Book Number: 978-1-63238-864-3 (Hardback)

Cataloging-in-Publication Data

Principles of mobile computing and communications / Mason Bell.
  p. cm.
Includes bibliographical references and index.
ISBN 978-1-63238-864-3
1. Mobile computing. 2. Mobile communication systems. 3. Electronic data processing.
4. Wireless communication systems. I. Bell, Mason.
QA76.59 .P75 2022
004--dc23

Contents

Permissions

Index

Preface

The human-computer interaction where the computer is typically designed to be transported during regular usage, is known as mobile computing. It allows the transmission of data, video and voice. The three aspects of mobile computing are mobile software, mobile communication and mobile hardware. Some of the main principles which lie behind mobile computing are portability, social interactivity, connectivity and individuality. Mobile computing makes use of primarily three different forms of wireless data connections. These are cellular data services, Wi-Fi connections and satellite internet access. Cellular data services, in turn, make use of different technologies like CDMA, GSM, EDGE and LTE. This book provides significant information of this discipline to help develop a good understanding of mobile computing and related fields. It includes contributions of experts and scientists which will provide innovative insights into this field. Those in search of information to further their knowledge will be greatly assisted by this book.

Given below is the chapter wise description of the book:

Chapter 1- Mobile computing deals with the transfer of data, voice and information over a network using a mobile. Mobile communication refers to the communication network that provides the cell with network coverage via a wireless medium. This is an introductory chapter which will briefly introduce about mobile computing and communication.

Chapter 2- Mobile device management is the type of security software that administers and manages mobile devices including smartphones, tablets, laptops, etc. Bring your own device, mobile application and content management, remote monitoring and management, etc. fall under its domain. This chapter has been carefully written to provide an easy understanding of mobile device management.

Chapter 3- Mobile network generations refer to the levels of evolution of mobile networks over a period of time. It includes first generation, second generation, third generation, fourth generation, fifth generation, LTE network, etc. The topics elaborated in this chapter will help in gaining a better perspective about these mobile network generations.

Chapter 4- Mobile communication network security is an essential component in mobile computing as it provides security of personal and business information stored in smartphones. Some of its aspects are GSM security, CDMA security, 3G security, 4G and LTE network security. This chapter discusses these aspects related to mobile communication network security in detail.

Chapter 5- Mobile communication standards and protocols make use of multiplexing for sending information. GSM architecture, code division multiple access, general packet radio service, next generation network, etc. are some of the concepts within it. This chapter closely examines these key concepts of mobile communication protocols to provide an extensive understanding of the subject.

Indeed, my job was extremely crucial and challenging as I had to ensure that every chapter is informative and structured in a student-friendly manner. I am thankful for the support provided by my family and colleagues during the completion of this book.

Mason Bell

Introduction to Mobile Computing and Communication

Mobile computing deals with the transfer of data, voice and information over a network using a mobile. Mobile communication refers to the communication network that provides the cell with network coverage via a wireless medium. This is an introductory chapter which will briefly introduce about mobile computing and communication.

Mobile Computing

Mobile Computing is a technology that allows transmission of data, via a computer, without having to be connected to a fixed physical link.

The term "Mobile computing" is used to describe the use of computing devices, which usually interact in some fashion with a central information system--while away from the normal, fixed workplace. Mobile computing technology enables the mobile worker to create, access, process, store and communicate information without being constrained to a single location. By extending the reach of an organization's fixed information system, mobile computing enables interaction with organizational personnel that were previously disconnected. It provides the continuous access to the wireless network services and the flexible communication between the people. It provides the real-time business to employee communication, enhanced customers interactions, and fastest communication between the individuals. The communication occurs with the real-time wireless connection. It provides the data, audio and video access to any user, any time with a wireless enable device.

The wireless network may be WLAN, Wi-Fi, GSM, CDMA, Wimax or GPRS. There are many companies that provide the mobile computing solutions on contract and pay as you go mobile broadband plans to the home users and businesses. 'The cell phones and laptops are the most commonly used mobile computing devices. It can be referred to the two main fields portable and mobility.

Computing: It can be used to check the email via the mobile phones, sending SMS, accessing internet and sending MMS. This technology has enabled the users to remain connected while on the move and it provides all the benefits of the computer network but without the cables. There are many companies that provide the mobile computing solutions to the home users and businesses. The portable device that uses this technology is the laptop computers.

Mobile computing devices can access any type of wireless network such as Wi-Fi, Wimax and wireless conventional network to access the internet and the network.

Mobile computing services can be provided for the specific purposes and its cost varies from company to company. Additionally, there are customized mobile computing solutions that are designed for the different commercial fields like health care, business, education, pharmaceutical, IT and service providers.

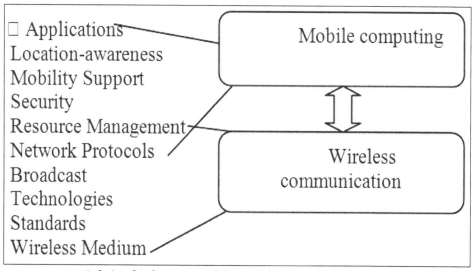

Relationship between mobile and wireless communication.

Principles of Mobile Computing

- Portability: Facilitates movement of device(s) within the mobile computing environment.

- Connectivity: Ability to continuously stay connected with minimal amount of lag/downtime, without being affected by movements of the connected nodes.

- Social Interactivity: Maintaining the connectivity to collaborate with other users, at least within the same environment.

- Individuality: Adapting the technology to suit individual needs.

- Portability: Devices/nodes connected within the mobile computing system should facilitate mobility.

- Connectivity: This defines the quality of service (QoS) of the network connectivity. In a mobile computing system, the network availability is expected to be maintained at a high level with the minimal amount of lag/downtime without being affected by the mobility of the connected nodes.

- Interactivity: The nodes belonging to a mobile computing system are connected with one another to communicate and collaborate through active transactions of data.

- Individuality: A portable device or a mobile node connected to a mobile network often denote an individual; a mobile computing system should be able to adopt the technology to cater the individual needs and also to obtain contextual information of each node.

Characteristics of Mobile Computing

Mobile computing is skilled using a grouping of component, system and applications package and a few kinds of communications medium. Potent mobile results have recently become potential due to the supply of particularly powerful and tiny computing devices, specialized package and improved telecommunication. A number of the characteristics of mobile computing is predicated on following:

Hardware: A short summary of the final forms of software, hardware, and communications averages that are ordinarily integrated to form mobile computing solutions are outlined. The characteristics of mobile computing hardware are outlined by the dimensions and kind issue, weight, micro chip, primary storage, external storage, screen size and kind, suggests that of input, suggests that of output, battery life, communications capabilities, expandability and sturdiness of the device.

Software: Mobile computers build use of a good type of system and application package. The primary common systems enclose and operative locations used on mobile computers include MS-DOS, Windows 95/98/NT, UNIX, android etc. These operative settings place capabilities from a discreet graphically- enhanced- pen- allowed DOS surroundings to the powerful potential of Windows NT. every operative system/environment has some kind of integrated development setting for application development. Most of the operative settings offer over one development environment choice for custom application development.

Communication: The flexibility of a mobile pc to speak in some fashion with a set system may be a process characteristic of mobile computing. The kind and accessibility of communication medium considerably impacts the kind of mobile computing application that may be created.

The means a mobile computer communicates with a set system are often categorized as: (a) connected (b) feeble connected (c) batch and (d) disconnected. The connected class implies an unceasingly accessible high-speed association. The flexibility to speak unceasingly, however at slow speeds, permits mobile computers to be feeble connected to the mounted system. A batch association means the mobile pc is not unceasingly accessible for communication with the mounted system. Within the batch mode, communication is established arbitrarily or sporadically to exchange and update info between the mobile pc and glued info systems. Mobile computers might operate in batch mode over communication mediums that are able of continuous operation, falling the wireless airtime and associated fees. Disconnected mobile computers permit users to enhance potency by creating calculations, storing contact info, keeping a schedule, and different non- communications directed tasks. This mode of operation is of very little interest as a result of the mobile device is unable of electronically interacting and replacing info with the mounted structure system. Exchange of data with a disconnected mobile computer will solely be accomplished by manually getting into information into the device or repeating from the device's screen and manually getting into the knowledge into the mounted system. This type of knowledge exchange is not any extra reasonable than mistreatment paper and is effectively nonexistent, since nearly all trendy mobile computing hardware is capable of some kind of native electronic knowledge communications. Knowledge Communications is that the exchange of knowledge mistreatment existing communication networks.

The term knowledge covers a good vary of applications as well as File Transfer, interrelation between Wide-Area-Networks (WAN), reproduction, email correspondence, access to the web and also the World Wide web (WWW).

Limitations of Mobile Computing

- Deficient Bandwidth: Mobile web access is usually slower than direct cable connections, victimisation technologies like GPRS and EDGE, and additional recently 3G networks. These networks are typically accessible at intervals vary of economic mobile phone towers. Higher speed wireless LANs are cheap however have terribly restricted vary.

- Security Standards: Once operating mobile, one relies on public networks, requiring careful use of Virtual personal Network (VPN). Security could be a major concern whereas regarding the mobile computing standards on the fleet. One will simply attack the VPN through an enormous variety of networks interconnected through the road.

- Power consumption: Once an influence outlet or moveable generator is not accessible, mobile computers should swear entirely on battery power. Combined with the compact size of the many mobile devices, this typically means that uncover priced batteries should be accustomed get the required battery life. Mobile computing should additionally look at Greener IT, in such the simplest way that it saves the ability or will increase the battery life.

- Transmission interferences: Weather, terrain, and also differ from the closest signal purpose will all interfere with signal response. Reception in tunnels, some structures, and rural areas is usually poor.

- Potential health hazards: Those that use mobile devices whereas driving are usually distracted from driving are so assumed additional probably to be concerned in traffic accidents. Cell phones could interfere with sensitive medical devices. There are allegations that mobile phone signals could cause health issues.

- Human interface with mechanism: Screens and keyboards are likely to be little, which can create them exhausting to use. Alternate input strategies like speech or handwriting recognition need coaching.

Applications of Mobile Computing

Mobile operating infrastructure will deliver real time business advantages, corporations of all sizes are walking up to the actual fact that they will improve productivity and increase profits by giving workers remote access to mission essential company IT system. The importance of Mobile Computers has been highlighted in several fields of that some are represented below:

- For Estate Agents: Estate agents will work either reception or call at the sector. With mobile computers they will be a lot of productive. They will acquire current property data by accessing multiple listing services, that they will do from home, workplace or automobile once out with purchasers. They will give shoppers with immediate feedback concerning specific homes or neighborhoods, and with quicker loan approvals, since applications is submitted on the spot. Therefore, mobile computers enable them to devote longer to purchasers.

- Emergency Services: Ability to receive data on the move is important wherever the emergency services are concerned. Data concerning the address, kind and different details of an event is sent quickly, via a Cellular Digital Packet data (CDPD) system using mobile computers, to 1 or many acceptable mobile units that are within the locality of the incident

- In courts: Defense counsels will take mobile computers in court. Once the opposing counsel references a case that they are not familiar, they'll use the computer to induce direct, period access to on-line legal information services, wherever they'll gather data on the case and connected precedents. Mobile computers enable immediate access to a wealth of data, creating individuals higher knowing and ready.

- In companies: Managers will use mobile computers in, say, and essential shows to major customers. They will access the most recent market share data. At a tiny low recess, they will revise the presentation to require advantage of this data. They will communicate with the workplace regarding doable new offers and decision conferences for discussing responds to the new proposals. Therefore, mobile computers will leverage competitive blessings.

- MasterCard Verification: At purpose of Sale (POS) terminals in retailers and supermarkets, once customers use credit cards for transactions, the communication is needed between the bank central pc and also the POS terminal, so as to result verification of the cardboard usage, will happen quickly and firmly over cellular channels using a mobile computer unit. This could speed up the dealings method and relieve congestion at the POS terminals.

Threats and Security Issues in Mobile Computing

Mobile computing brings with it threats to the user and to the corporate environment. From personal information to corporate data, mobile devices are used for a wide variety of tasks by individuals and companies. Mobile devices have added a new threat to the corporate landscape as they have introduced the concept of bring your own device. While this is not necessarily an entirely new concept, the wide acceptance of bring your own device with mobile devices has created a paradigm shift, where the security and safety of the device is not necessarily to protect the corporate data, but to keep the personal data out of the hands of corporate management.

Data Loss from lost, stolen, or decommissioned devices: By their nature, mobile devices are with us everywhere we go. The information accessed through the device means that theft or loss of a mobile device has immediate consequences. Additionally, weak password access, no passwords, and little or no encryption can lead to data leakage on the devices. Users may also sell or discard devices without understanding the risk to their data. The threat level from data loss is high, as it occurs frequently and is a top concern across executives and IT admin.

Information stealing mobile malware: Android devices, in particular, offer many options for application downloads and installations. Unlike iOS devices, which need to be jail broken, Android users can easily opt to download and install apps from third-party marketplaces other than Google's official Play Store marketplace. To date, the majority of malicious code distributed for Android has been disseminated through third-party app stores. Most of the malware distributed through third-party stores has been designed to steal data from the host device. This threat level is high, as Android malware in particular is becoming a more popular attack surface for criminals who traditionally have used PCs as their platforms. Data Loss and data leakage through poorly written third-party applications: Applications for smart phones and tablets have grown exponentially on iOS and Android. Although the main marketplaces have security checks, certain data collection processes are of questionable necessity; all too often, applications either ask for too much access to data or simply gather more data than they need or otherwise advertise. This is a mid-level threat. Although data loss and leaking through poorly written applications happens across mobile operating systems.

Vulnerabilities within devices, OS, design, and third-party applications: Mobile hardware, OS, applications and third-party apps contain defects (vulnerabilities) and are susceptible to ex-filtration and/or injection of data and/or malicious code (exploits). The unique ecosystem inherent in mobile devices provides a specialized array of security concerns to hardware, OS, and application developers, as mobile devices increasingly contain all of the functionalities attributed to desktop computing, with the addition of cellular communication abilities. This is a mid-level threat; although the possibility is high, the number of exploits is not.

Unsecured WiFi, network access, and rogue access points: This has increased the attack surface for users who connect to these networks. In the last year, there has been a proliferation of attacks on hotel networks, a skyrocketing number of open rogue access points installed, and the reporting of eavesdropping cases. This threat level is high. Increased access to public WiFi, along with increased use of mobile devices, creates a heightened opportunity for abuse of this connection.

Unsecured or rogue marketplaces: Android users can easily opt to download and install apps from third-party marketplaces other than Google's official Play Store marketplace. To date, the majority of malicious code distributed for Android has been distributed through third-party app stores. This threat level is high: Android malware in particular is being distributed through these marketplaces more and more frequently.

Insufficient management tools, capabilities, and access to APIs (includes personas): Granting users and developers access to a device's low-level functions is a double-edged sword, as attackers, in theory, could also gain access to those functions. However, a lack of access to system-level functions to trusted developers could lead to insufficient security. Additionally, with most smart phone and tablet operating systems today, there is little, if any, guest access or user status. Thus, all usage is in the context of the admin, thereby providing excessive access in many instances. This is a mid-level threat.

NFC and proximity-based hacking: Near-field communication (NFC) allows mobile devices to communicate with other devices through short-range wireless technology. NFC technology has been used in payment transactions, social media, coupon delivery, and contact information sharing. Due to the information value being transmitted, this is likely to be a target of attackers in the future. The threat level is low, as the threat is still in the proof-of-concept phase.

Security Countermeasures

Secure mobile computing is critical in the development of any application of wireless networks.

Security Requirements

Similar to traditional networks, the goals of securing mobile computing can be defined by the following attributes: Availability, confidentiality, integrity, authenticity and non-repudiation.

- Availability ensures that the intended network services are available to the intended parties when needed.

- Confidentiality ensures that the transmitted information can only be accessed by the intended receivers and is never disclosed to unauthorized entities.

- Authenticity allows a user to ensure the identity of the entity it is communicating with. Without authentication, an adversary can masquerade a legitimate user, thus gaining unauthorized access to resource and sensitive information and interfering with the operation of users.

- Integrity guarantees that information is never corrupted during transmission. Only the authorized parties are able to modify it.

- Non-repudiation ensures that an entity can prove the transmission or reception of information by another entity, i.e. sender/receiver cannot falsely deny having received or sent certain data.

- In ad hoc networks, mobile hosts are not bound to any centralized control like base stations or access points. They are roaming independently and are able to move freely with an arbitrary speed and direction. Thus, the topology of the network may change randomly and frequently. In such a network, the information transfer is implemented in a multi-hop fashion, i.e., each node acts not only as a host, but also as a router, forwarding packets for those nodes that are not in direct transmission range with each other. By nature, an ad hoc network is a highly dynamic self-organizing network with scarce channels. Besides these security risks, ad hoc networks are prone to more security threats due to their difference from conventional infrastructure-based wireless networks.

- The lack of pre-fixed Infrastructure means there is no centralized control for the network services. The network functions by cooperative participation of all nodes in a distributed fashion. The decentralized decision making is prone to the attacks that are designed to break the cooperative algorithms. A malicious user could simply block or modify the traffic traversing it by refusing to cooperate and break the cooperative algorithms. Moreover, since there are no trusted entities that can calculate and distribute the secure keys, the traditional key management scheme cannot be applied directly.

- Dynamically Changing Topology aids the attackers to update routing information maliciously by pretending this to be legitimate topological change. In most routing protocols for ad hoc networks, nodes exchange information about the topology of the network so that the routes could be established between communicating nodes. Any intruder can maliciously give incorrect updating information. For instance, DoS attack can be easily launched if a malicious node floods the network with spurious routing messages. The other nodes may unknowingly propagate the messages.

- Energy Consumption Attack is more serious as each mobile node also forwards packets for other nodes. An attacker can easily send some old messages to a node, aiming to overload the network and deplete the node's resources. More seriously, an attack can create a rushing attack by sending many routing request packets with high frequency, in an attempt to keep other nodes busy with the route discovery process, so the network service cannot be achieved by other legitimate nodes.

- Node Selfishness is a specific security issue to ad hoc network. Since routing and network management are carried by all available nodes in ad hoc networks, some nodes may selfishly deny the routing request from other nodes to save their own resources (e.g., battery power, memory, CPU).

Security Issues involved in Mobile Computing

- Mobile security or mobile phone security has become increasingly important in mobile computing. It is of particular concern as it relates to the security of personal information now stored on the smart phone. More and more users and businesses use smart phones as communication tools but also as a means of planning and organizing their work and private life. Within companies, these technologies are causing profound changes in the organization of information systems and therefore they have become the source of new risks. Indeed, smart phones collect and compile an increasing amount of sensitive information to which access must be controlled to protect the privacy of the user and the intellectual property of the company.

Security Vulnerabilities of a General Mobile Computing System

	Mobile Units	Over the air	Wired hosts
Physical vulnerabilities	Small size and weight, portability, exposure in hostile places.	Random happenings that easy affect wireless communications.	Different locations
Natural vulnerabilities	Exposure in outdoor environmental conditions.	Affected from weather situations, hand-offs between cells.	Unknown boundaries many points to attack.
H/W and S/W vulnerabilities	Not enough hardware controls and resources.		Heterogeneity, shared use of resources.
Communications vulnerabilities	Dependence on the communication infrastructure.	Broadcasting	
Human vulnerabilities	Away from technical support and management, lack of attention.	Unlimited capability physical access.	

- All smart phones, as computers, are preferred targets of attacks. These attacks exploit weaknesses related to smart phones that can come from means of communication like SMS, MMS, WIFI NETWORKS. There are also attacks that exploit software vulnerabilities from both the web browser and operating system.

- Different security counter-measures are being developed and applied to smart phones, from security in different layers of software to the dissemination of information to end users. There are good practices to be observed at all levels, from design to use, through the development of operating systems, software layers, and downloadable apps.

- One of the key issues of these being, confidentiality and authentication, where the user must be protected from unauthorized eavesdropping. The goal of authentication protocol is to check the identity of other users or network centers before providing access to the confidential information on the user side. When designing any security protocol, there are certain conditions that need to be considered. Firstly, the low computational power of the mobile users and secondly, the low bandwidth available. Therefore, it is important to design the security protocols so as to minimize, the number of message exchanges and the message size. a few authentication protocols that were proposed to provide security between the users and the network. These protocols are based on the use of certificates, which are built on the concept of security keys (cryptography). Another protocol that is discussed in this paper is the Kilo Byte Secure Socket Layer (KSSL) protocol, which is an extension of Secure Socket Layer (SSL) protocol used for wired networks.

Classification of Security Attacks

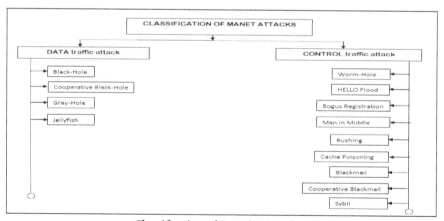

Classification of Security Attacks.

DATA Traffic Attack

DATA traffic attack deals either in nodes dropping data packets passing through them or in delaying of forwarding of the data packets. Some types of attacks choose victim packets for dropping while some of them drop all of them irrespective of sender nodes. This may highly degrade the quality of service and increases end to end delay. This also causes significant loss of important data. For e.g., a 100Mbps wireless link can behave as 1Mbps connection. Moreover, unless there is a redundant path around the erratic node, some of the nodes can be unreachable from each other altogether.

- Black-Hole Attack: In this attack, a malicious node acts like a Black hole, dropping all data packets passing through it as like matter and energy disappears from our universe in a black hole. If the attacking node is a connecting node of two connecting components of that network, then it effectively separates the network in to two disconnected components the Black-Hole node separates the network into two parts. Few strategies to mitigate the problem: (i) collecting multiple RREP messages (from more than two nodes) and thus hoping multiple redundant paths to the destination node and then buffering the packets until a safe route is found. (ii) Maintaining a table in each node with previous sequence number in increasing order. Each node before forwarding packets increases the sequence number. The sender node broadcasts RREQ to its neighbors and once this RREQ reaches the destination, it replies with a RREP with last packet sequence number. If the intermediate node finds that RREP contains a wrong sequence number, it understands that somewhere something went wrong.

- Cooperative Black-Hole Attack: This attack is similar to Black-Hole attack, but more than one malicious node tries to disrupt the network simultaneously. It is one of the most severe DATA traffic attack and can totally disrupt the operation of an Ad Hoc network. Mostly the only solution becomes finding alternating route to the destination, if at all exists. Detection method is similar to ordinary Black-Hole attack. In addition another solution is securing routing and node discovery in MANET by any suitable protocol such as SAODV, SNRP, SND, SRDP etc. Since each node is already trusted, black hole node should not be appearing in the network.

- Gray-Hole Attack: Gray-Hole attack has its own characteristic behavior. It too drops DATA packets, but node's malicious activity is limited to certain conditions or trigger. Two most common type of behavior: (i) Node dependent attack – drops DATA packets destined towards a certain victim node or coming from certain node, while for other nodes it behaves normally by routing DATA packets to the destination nodes correctly. (ii) Time dependent attack – drops DATA packets based on some predetermined/trigger time while behaving normally during the other instances. Detecting this behaviorist attack is very difficult unless there exist a system wide detection algorithm, which takes care of all the nodes performance in the network. Sometimes nodes can interact with each other and can advise malicious nodes existence to other friendly nodes. Approach is similar to Black-Hole attack where sequence number feedback might detect some Gray-Hole attack. If multiple paths exist between sender and destination then buffering packets with proper acknowledgement might detect active Gray-Hole attack in progress. But dormant or triggered attack is difficult to detect with this approach.

- Jellyfish Attack: Jellyfish attack is somewhat different from Black-Hole & Gray-Hole attack. Instead of blindly dropping the data packets, it delays them before finally delivering them. It may even scramble the order of packets in which they are received and sends it in random order. This disrupts the normal flow control mechanism used by nodes for reliable transmission. Jellyfish attack can result in significant end to end delay and thereby degrading QoS.

Control Traffic Attack

Mobile Ad-Hoc Network (MANET) is inherently vulnerable to attack due to its fundamental characteristics, such as open medium, distributed nodes, autonomy of nodes participation in network (nodes can join and leave the network on its will), lack of centralized authority which can enforce security on the network, distributed co-ordination and cooperation. The existing routing protocols cannot be used in MANET due to these reasons. Many of the routing protocols devised for use in MANET have their individual characteristic and rules. Two of the most widely used routing protocols is AdHoc On Demand Distance Vector routing protocol (AODV), which relies on individual node's cooperation in establishing a valid routing table and Dynamic MANET OnDemand (DYMO), which is a fast light weight routing protocol devised for multi hop networks. But each of them is based on trust on nodes participating in network. The first step in any successful attack requires the node to be part of that network. As there is no constraint in joining the network, malicious node can join and disrupts the network by hijacking the routing tables or bypassing valid routes. It can also eavesdrop on the network if the node can establish itself as the shortest route to any destination by exploiting the unsecure routing protocols. Therefore it is of utmost importance that the routing protocol should be as much secure as it can be.

- Worm Hole Attack: Worm hole, in cosmological term, connects two distant points in space via a shortcut route. In the same way in MANET also one or more attacking node can disrupt routing by short-circuiting the network, thereby disrupting usual flow of packets. If this link becomes the lowest cost path to the destination then these malicious nodes will always be chosen while sending packets to that destination. The attacking node there have been few proposals recently to protect networks from worm-hole attack: Geographical leashes & temporal leashes: A leash is added to each packet in order to restrict the distance the packets are allowed to travel. A leash is associated with each hop. Thus, each transmission of a packet requires a new leash. A geographical leash is intended to limit the distance between the transmitter and the receiver of a packet. A temporal leash provides an upper bound on the lifetime of a packet. Using directional antenna: Using directional antenna restricts the direction of signal propagation through air. This is one of the crude ways of limiting packet dispersion.

- HELLO Flood Attack: The attacker node floods the network with a high quality route with a powerful transmitter. So, every node can forward their packets towards this node hoping it to be a better route to destination. Some can forward packets for those destinations which are out of the reach of the attacker node. A single high power transmitter can convince that all the nodes are his neighbor. The attacker node need not generate a legitimate traffic; it can just perform a selective replay attack as its power overwhelms other transceivers.

- Bogus Registration Attack: A Bogus registration attack is an active attack in which an attacker disguises itself as another node either by sending stolen beacon or generating such false beacons to register himself with a node as a neighbor. Once registered, it can snoop transmitted packets or may disrupt the network altogether. But this type of attack is difficult to achieve as the attacker needs to intimately know the masquerading nodes identity and network topology. Encrypting packets before sending and secure authentication in route discovery (SRDP, SND, SNRP, ARAN, etc) will limit the severity of attack to some extent as attacker node has no previous knowledge of encryption method.

- Man in Middle Attack: In Man in Middle attack, the attacker node creeps into a valid route and tries to sniff packets flowing through it. To perform man in middle attack, the attacker first needs to be part of that route. It can do that by either temporarily disrupting the route by deregistering a node by sending malicious disassociation beacon captured previously or registering itself in next route timeout event. One way of protecting packets flowing through MANET from prying eyes is encrypting each packet. Though key distribution becomes a security issue.

- Rushing Attack: In AODV or related protocol, each node before transmitting its data, first establishes a valid route to destination. Sender node broadcasts a RREQ (route request) message in neighborhood and valid routes replies with RREP (route reply) with proper route information. Some of the protocols use duplicate suppression mechanism to limit the route request and reply chatter in the network. Rushing attack exploits this duplicate suppression mechanism. Rushing attacker quickly forwards with a malicious RREP on behalf of some other node skipping any proper processing. Due to duplicate suppression, actual valid RREP message from valid node will be discarded and consequently the attacking node becomes part of the route. In rushing attack, attacker node does send packets to proper node after its own filtering is done, so from outside the network behaves normally as if nothing happened. But it might increase the delay in packet delivering to destination node.

- Cache Poisoning Attack: Generally in AODV, each node keeps few of its most recent transmission routes until timeout occurs for each entry. So each route lingers for some time in node's memory. If some malicious node performs a routing attack then they will stay in node's route table until timeout occurs or a better route is found. An attacker node can advertise a zero metric to all of its destinations. Such route will not be overwritten unless timeout occurs. It can even advertise itself as a route to a distant node which is out of its reach. Once it becomes a part of the route, the attacker node can perform its malicious activity. Effect of Cache poisoning can be limited by either adding boundary leashes or by token authentication. Also each node can maintain its friend-foe list based on historical statistics of neighboring nodes performance.

- Blackmailing and Co-operative Blackmailing Attack: In a blackmailing attack or more effectively co-operative blackmailing attack, attacker nodes accuse an innocent node as harmful node. This attack can effectively be done on those distributed protocols that establish a good and bad node list based on review of participating nodes in MANET. Few of the protocols tries to make them more secure by using majority voting principle, but still if sufficient no. of attacker nodes become part of the MANET it can bypass that security also. Another generic method of this attack will be, sending invalid RREP messages with advertising an unnecessarily high cost to certain nodes.

- Sybil Attack: Sybil attack manifests itself by faking multiple identities by pretending to be consisting of multiple nodes in the network. So one single node can assume the role of multiple nodes and can monitor or hamper multiple nodes at a time. If Sybil attack is performed over a blackmailing attack, then level of disruption can be quite high. Success in Sybil attack depends on how the identities are generated in the system.

Mobile Communication

Mobile Communication is the use of technology that allows us to communicate with others in different locations without the use of any physical connection (wires or cables). Mobile communication makes our life easier, and it saves time and effort.

A mobile phone (also called mobile cellular network, cell phone or hand phone) is an example of mobile communication (wireless communication). It is an electric device used for full duplex two way radio telecommunication over a cellular network of base stations known as cell site.

Features of Mobile Communication

The following are the features of mobile communication:

- High capacity load balancing: Each wired or wireless infrastructure must incorporate high capacity load balancing. High capacity load balancing means, when one access point is overloaded, the system will actively shift users from one access point to another depending on the capacity which is available.

- Scalability: The growth in popularity of new wireless devices continuously increasing day by day. The wireless networks have the ability to start small if necessary, but expand in terms of coverage and capacity as needed - without having to overhaul or build an entirely new network.

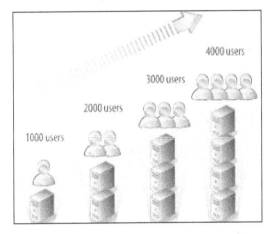

- Network management system: Now a day, wireless networks are much more complex and may consist of hundreds or even thousands of access points, firewalls, switches, managed power and various other components. The wireless networks have a smarter way of managing the entire network from a centralized point.

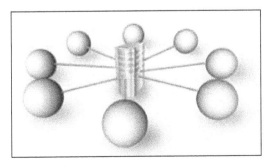

- Role based access control: Role based access control (RBAC) allows you to assign roles based on what, who, where, when and how a user or device is trying to access your network. Once the end user or role of the devices is defined, access control policies or rules can be enforced.

○   Indoor as well as outdoor coverage options: It is important that your wireless system has the capability of adding indoor coverage as well as outdoor coverage.

○   Network access control: Network access control can also be called as mobile device registration. It is essential to have a secure registration. Network access control (NAC) controls the role of the user and enforces policies. NAC can allow your users to register themselves to the network. It is a helpful feature that enhances the user experience.

○   Mobile device management: Suppose, many mobile devices are accessing your wireless network; now think about the thousands of applications are running on those mobile devices. How do you plan on managing all of these devices and their applications, especially as devices come and go from your business? Mobile device management can provide control of how you will manage access to programs and applications. Even you can remotely wipe the device if it is lost or stolen.

•   Roaming: You don't need to worry about dropped connections, slower speeds or any disruption in service as you move throughout your office or even from building to building wireless needs to be mobile first. Roaming allows your end-users to successfully move from one access point to another without ever noticing a dip in a performance. For example, allowing a student to check their mail as they walk from one class to the next.

•   Redundancy: The level or amount of redundancy your wireless system requires depends on your specific environment and needs.

- For example: A hospital environment will need a higher level of redundancy than a coffee shop. However, at the end of the day, they both need to have a backup plan in place.

- Proper Security means using the right firewall: The backbone of the system is your network firewall. With the right firewall in place you will be able to:

  - See and control both your applications and end users.

  - Create the right balance between security and performance.

  - Reduce the complexity with:

    - Antivirus protection.

    - Deep Packet Inspection (DPI).

    - Application filtering.

  - Protect your network and end users against known and unknown threads including:

    - Zero- day.

    - Encrypted malware.

    - Ransomware.

    - Malicious botnets.

- Switching: Basically, a network switch is the traffic cop of your wireless network which making sure that everyone and every device gets to where they need to go. Switching is an essential part of every fast, secure wireless network for several reasons:

  - It helps the traffic on your network flow more efficiently.

  - It minimizes unnecessary traffic.

  - It creates a better user experience by ensuring your traffic is going to the right places.

Advantages of Mobile Communication

There are following advantages of mobile communication:

- Flexibility: Wireless communication enables the people to communicate with each other regardless of location. There is no need to be in an office or some telephone booth in order to pass and receive messages.

- Cost effectiveness: In wireless communication, there is no need of any physical infrastructure (Wires or cables) or maintenance practice. Hence, the cost is reduced.

- Speed: Improvements can also be seen in speed. The network connectivity or the accessibility was much improved in accuracy and speed.

- Accessibility: With the help of wireless technology easy accessibility to the remote areas is possible. For example, in rural areas, online education is now possible. Educators or students no longer need to travel to far-flung areas to teach their lessons.

- Constant connectivity: Constant connectivity ensures that people can respond to emergencies relatively quickly. For example, a wireless device like mobile can ensure you a constant connectivity though you move from place to place or while you travel, whereas a wired landline can't.

Cellular Communication has become an important part of our daily life. Besides using cell phones for voice communication, we are now able to access the Internet, conduct monetary transactions, send text messages etc. using our cell phones, and new services continue to be added. However, the wireless medium has certain limitations over the wired medium such as open access, limited bandwidth and systems complexity. These limitations make it difficult although possible to provide security features such as authentication, integrity and confidentiality. The current generation of 3G networks have a packet switched core which is connected to external networks such as the Internet making it vulnerable to new types of attacks such as denial of service, viruses, worms etc. that have been used against the Internet.

Mobile Hardware

Mobile hardware includes mobile devices or device components that receive or access the service of mobility. They would range from portable laptops, smartphones, tablet Pc's, Personal Digital Assistants.

These devices will have a receptor medium that is capable of sensing and receiving signals. These devices are configured to operate in full- duplex, whereby they are capable of sending and receiving signals at the same time. They don't have to wait until one device has finished communicating for the other device to initiate communications.

Above mentioned devices use an existing and established network to operate on. In most cases, it would be a wireless network.

Mobile Software

Mobile software is the actual program that runs on the mobile hardware. It deals with the characteristics and requirements of mobile applications. This is the engine of the mobile device. In other terms, it is the operating system of the appliance. It's the essential component that operates the mobile device.

Since portability is the main factor, this type of computing ensures that users are not tied or pinned to a single physical location, but are able to operate from anywhere. It incorporates all aspects of wireless communications.

Mobile Computing Device

A mobile computing device is any device that is created using mobile components, such as mobile hardware and software. Mobile computing devices are portable devices capable of operating, executing and providing services and applications like a typical computing device.

Mobile computing devices also may be known as portable computing devices or handheld computing devices.

Mobile computing devices are generally modern-day handheld devices that have the hardware and software required to execute typical desktop and Web applications. Mobile computing devices have similar hardware and software components as those used in personal computers, such as processors, random memory and storage, Wi-Fi, and a base operating system. However, they differ from PCS in that they are built specifically for mobile architecture and to enable portability.

Among the common examples of mobile computing devices is a tablet PC, which has a built-in processor, memory and operating system (OS), and executes most applications built for a comparable PC.

Mobile computer devices are designed to be portable, often to fit on your lap, in the palm of your hand or in your pocket. With some mobile devices, you can do many of the things you do with a desktop computer while you are away from home or traveling. Features in mobile computer devices include batteries, video camera, camera, voice recorder and music player. Laptop computers, tablets, smartphones, e-readers and hand-held gaming devices are five types of mobile technology used to connect to the Internet and communicate with others.

Laptop Computers

Laptop computers are personal computers that are easy to carry and use in various locations. Many laptops on the market are designed to offer you all the functionality of a desktop computer, which means you can run the same software and open the same types of files.

The laptop has an all-in-one design with built-in touchpad, keyboard, monitor and speakers. Laptops also offer you the option of connecting to a larger monitor, regular mouse and other peripherals. This feature means you can turn a laptop into a desktop computer, but one you can disconnect from the peripherals and carry with you wherever you go.

Most laptops have the same types of ports desktop computers have -- USB, HDMI and Firewire -- although there are usually fewer of them to save space. However, some laptop ports are different from desktop computers and may require an adapter to use them. For example, the monitor port on a laptop is a Mini DisplayPort, which is smaller than the DisplayPort on a desktop computer.

Tablets

Tablets are also designed to offer portability. However, they provide you with a computing experience different from laptops with the biggest difference being that tablets do not have a touchpad or keyboard. Instead, the touch screen offers a virtual keyboard you use to input text, while your finger replaces the mouse as a pointer.

Tablets are bigger than a smartphone and smaller than a laptop. Like the smartphone, you can browse the Internet, carry out videoconferences, stay connected through email, read e-books, play games, watch movies, share photos and listen to music with the tablet.

Basic features of tablet computers include:

- Mobile OS: Tablets run on mobile operating systems different from their desktop counterparts. Examples include Windows, iOS and Android.

- Solid-state drives: Tablets use solid-state drives, which are faster and more durable than hard disk drives.

- Wi-Fi: Because tablets are optimized for Internet use, they have built-in Wi-Fi.

Smartphones

A smartphone is a powerful mobile phone capable of running applications in addition to providing with phone service. These devices have most of the features available on tablets along with cellular Internet connectivity. Cell phone companies offer data plans that offer you Internet access anywhere with coverage.

E-readers

E-readers, or e-book readers, resemble tablet computers, but that they are mainly designed for reading digital and downloadable documents. E-readers have either an LCD or e-ink display.

- LCD Display: This is the same screen found on laptops and tablet computers. This type of screen is suitable for viewing books and magazines with photos because the LCD screen can display colors.

- E-Ink Display: E-ink is short for electronic ink and usually displays in black and white. It is designed to offer you the look of an actual page in a book. Unlike the LCD display, the e-ink version is not backlit, so text is readable even outdoors in full sun. E-ink displays offer a reading experience with less eyestrain.

Handheld Gaming Devices

Handheld gaming devices are portable, lightweight video game consoles that have built-in game controls, screen and speakers. With a handheld gaming console, you can play your favorite console games wherever you are, whether on the move or while someone else is watching the TV.

Basic features of handheld gaming devices such as Nintendo 3DS and PS Vita include:

- Online access to free and paid games.

- Access to online movies, TV shows.

- Social media apps.

- Web browsing.

- Online and local multiplayer support.

References

- Mobile-computing-concept-and-principles: datasagar.com, Retrieved 25 January, 2019

- Mobile-communication-introduction: javatpoint.com, Retrieved 02 March, 2019

- Mobile-computing-overview, mobile-computing: tutorialspoint.com, Retrieved 23 May, 2019

- Mobile-computing-device-mcd- 8270: techopedia.com, Retrieved 16 April, 2019

- Types-of-mobile-computer-devices: techwalla.com, Retrieved 19 July, 2019

- "Wireless communications security," Boston: Artech House, 2006

- Mobile Telecommunications factbook. New York: McGraw-Hill

Mobile Device Management

Mobile device management is the type of security software that administers and manages mobile devices including smartphones, tablets, laptops, etc. Bring your own device, mobile application and content management, remote monitoring and management, etc. fall under its domain. This chapter has been carefully written to provide an easy understanding of mobile device management.

Mobile device management (MDM) refers to the control of one or more mobile devices through various types of access control and monitoring technologies. This term is commonly related to enterprise use of mobile devices, where it is important for businesses to both allow for effective mobile device use, and protect sensitive data from unauthorized access.

Over time, mobile device management has advanced to include newer methods like remote server controls, more versatility in managing groups of diverse providers, and even software-as-a-service implemented controls. In older systems, users may have had to install a SIM card in a device to get access to internal systems, whereas newer MDM systems often operate through an over-the-air method. MDM has adapted to fit the needs of enterprises managing specific kinds of systems related to tech trends around smartphone and mobile device use.

One of the most prominent of these is bring your own device (BYOD). With BYOD, employers and employees have been able to share mobile use with flexible arrangements where a personal device can be used for business. Modern mobile device management adds security and scalability to these kinds of uses, which can require some complex access architectures and customized engineering.

The field of mobile device management has continued to evolve as more professionals are using laptops and smartphones to work. This has increased the need for solutions that allow employees to access information wherever they are and at any time. Early solutions focused solely on devices, and lacked application and content management; today, they are now growing into broader EMM solutions to better capture and serve the mobile opportunity.

Current EMM suites consist of policy- and configuration-management tools that are coupled with a management overlay for applications and content that's intended for mobile devices, which are smartphone-OS specific. IT organizations and service providers use EMM suites to deliver IT support to mobile end users and to maintain security policies.

Modern EMM suites provide the following core functions:

- Hardware inventory.

- Application inventory.

- OS configuration management.

- Mobile app deployment, updating and removal.

- Mobile app configuration and policy management.

- Remote view and control for troubleshooting.

- Execute remote actions, such as remote wipe.

- Mobile content management.

Key elements of an MDM solution include:

- Asset management, which includes multi-platform support for companies to apply custom organizational policies to enterprise mobility and BYO device use in the corporate network. Asset management might monitor and control how the devices can be used as well as enforce company policy across all enrolled devices, multiple platforms, and operating system versions.

- Configurations management, which can identify, control, and manage hardware and software settings based on geographic regions, user profiles, and identity.

- Risk management, audits, and reporting, which monitors device activity and reports anomalous behavior to limit issues such as unauthorize access of corporate network or data transfers.

- Software updates and distribution, which can remotely control applications, software and OS updates, and licenses across multiple devices.

- Profile management, which allows management of policies and settings to specific groups of end users based on specific profiles.

- Identity and access management, which ensures that the device, data, network connection, and services are provided to appropriate authorized users.

- Applications management, which distributes, manages settings, and black- and whitelists apps and software functionality.

- Enterprise app stores, which maintain a library of apps and services dedicated for corporate use that are available to authorized end-users.

- Bandwidth optimization, which manages bandwidth usage at the device and application level.

- Data security, which ensures that data is accessed, transferred, and utilized in accordance with organizational policies. For instance, in event of device theft or loss, data stored on the device can be wiped out remotely.

- Content management, which synchronizes and secures business information across multiple devices.

- Tech support, which includes dedicated remote technology support can be provided remotely.

Mobile Device Management: Complexity and Challenges

Growing interest in BYOD devices and mobility initiatives have fueled mass adoption of smartphones and tablets in the workplace. The promise of employee productivity, however, has not always matched the resulting business value. Potential gains in workforce productivity are often overshadowed by the challenge of managing a scalable pool of less-secure mobile devices, amid growing security threats and the dynamic technology landscape.

While enterprise mobility is a business reality that organizations must embrace, managing endpoint devices is not limited to defining and enforcing static policies. Furthermore, managing mobile devices and the expected enterprise mobility capabilities goes beyond purchasing and deploying MDM solutions.

Consider the case of mobile device fragmentation. As the device market grows, employees can choose devices from multiple brands, platforms, and software versions—which companies must then support and manage. Unforeseen security issues facing a specific platform and software version must be approached proactively. Yet, risk management must not compromise end-user convenience and preferences in terms of which device they can use and how they can use it for office tasks.

Another complication is that MDM features vary across vendors. A standalone MDM solution from a specific vendor may satisfy the device management needs of the organization today, but the changing technology and business circumstances may necessitate additional investments into changing EMM capabilities. Therefore, device management is not just a technology problem, but a strategic challenge that every company must manage with their own best practices, based on their unique requirements and the future landscape of technology.

MDM solutions are designed to enhance visibility and control into an end user's mobile device activity. But, excessive tracking of mobile device activity could compromise end-user privacy. For example, an MDM may track real-time location, browsing activity, information that reveals personal information and usage habits of employees beyond the device management and security needs of the employer.

Best Practices for Mobile Device Management

With the rapid proliferation of BYOD devices connecting to the corporate network, organizations must enforce device management controls without compromising the security posture of the business or the privacy and convenience of end users. To achieve a balance between both objectives, organizations can adopt the following best practices:

- Implement policies before deploying an MDM solution: The right set of policies should be established to meet the unique technical and business needs of the organization before deploying an MDM solution.

- Make device enrollment to MDM solutions easy and convenient: Ensure that no BYOD device goes under the radar, especially because of difficult or insufficient enrollment procedures or platform support.

- Establish self-service capabilities: End user self-service is crucial in maintaining compliance with MDM solutions. Self-service capabilities can include remote data wipe-out, password reset, and lost device tracking.

- Ensure up-to-date MDM versions: Push configuration changes, patch installations, and install software updates as soon as required and made available. A BYOD device running vulnerable outdated software is a security incident waiting to happen.

- Protect end-user privacy: This will become key to ensuring end users continue compliance. Protect employee privacy by restricting data collection to a bare minimum and establishing procedures to eliminate misuse of personal employee information while still aligning with the company's technical and business needs.

- Deploy containment technologies: These can separate corporate apps, data, and MDM controls from the personal use of a BYO device. With such containment in place, the MDM rules and features will only apply when the BYO device engages in corporate use.

- Monitor devices for specific activities or situations: Monitor devices for anomalous activities or under optimized data usage.

Components of Mobile Device Management

There are two broad branches of components in a successful implementation of a Mobile Device Management strategy viz. Standard Mobile Device Management components and Remote Mobile Device Management components.

Standard Mobile Device Management Components

Standard Mobile Device Management components consist of the infrastructure and software designed and operated to regulate the use of personal mobile devices within the workplace. They also include the key command central components that help oversee the broad implementation of a control strategy.

Central Management Console

The Central Management Console is the brain and hub of the Mobile Device Management network. It is used by the administrator to monitor the status, output and data usage of all the personal mobile devices that have been registered for usage by employees. This includes an interface that provides basic performance data and statistics that are measurable by maintaining connectivity between the device and the central workplace server via the usage of a local network. This hub can display various parameters while the device is in usage so as to monitor and optimise workflow.

Registry

The registry offers various setup and enrolment services to the user which aid in organisation of the workforce and the workflow. An example of such a service would be enrolment by invitation whereby groups of employees can create a virtual network that enables communication, sharing of files and upload, download and synchronising of data between users. This helps teams keep track of their work progress, be aware of deadlines, briefs and restrictions as well as being able to constantly keep updating each other about the state of work. Such a service could also provide a window for an overseeing authority to monitor the quality of work and make performance reports.

Asset Management Module

The Asset Management Module allows for the complete management of device inventory. This includes any and all work related files, information or any form of data holders so as to ensure compliance with standards and practices and to avoid any theft or unsolicited access to sensitive or unorganised data.

Security Management Module

The Security Management Module allows for a complete safety net that protects employee mobile device usage from threats such as unsafe or inappropriate website URLs, phishing risks or fraudulent transactions. This module also allows for the admin accessing a comprehensive report on employee network usage and allows for the blockage of unauthorised websites as well as a spam filter. Broadly it can be broken up into three categories:

Secure E-mail

MDM allows for industry standard security encryption that can make the sending and receiving of email as well as file transfers via email a procedure that is safer than using a standard email client. It also offers secure options for contact management and provides superior interconnectivity between departments.

Secure Browsing

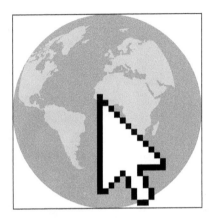

Secure browsing protects the user from malicious websites by restricting or blocking access to them. It provides for the easiest way to monitor and regulate employee browsing activity in order to ensure risk free Internet access.

Application Catalogue

The presence of an organised catalogue can help in the whitelisting or blacklisting of mobile applications to be accessed by the user. It can also centralise the process of updating of applications as well as their distribution and management. It also provides for special permissions to allow certain applications to run on kiosk mode so as to better regulate their use.

Kiosk Mode

Kiosk Mode is a special form of operation of a device wherein certain options to alter system components and/or functionalities are limited either by restriction or denial of access. Fundamentally, the device operates as a "kiosk" where the user has only a limited range of options of functionality presented on the user interface. This enables the device to be restricted to only being viable for professional use. It also provides for increased security as the data can only be accessed with the proper administrator credentials. It prevents misuse of the device that can hamper productivity or quality of work output.

Automation Module

The Automation Module allows for the condensing of a package of features and functionality into an automated process that saves both human work time as well as reduces error. This includes service and support requests (for example, a ticketing system that catalogues a user's support history. This would allow for faster diagnosis of problems based on usage history as well as streamline any requests for support in order to optimise the resolution process), usage monitoring and data allowance as well as policy enforcement. Often times this provides a timely notification on issues of data usage, lag, or notifications of expected downtime and repair processes.

Self Service Module

The Self Service feature will help a user skip the process of raising a support request if the problem is one that can be solved by the user themselves. This is particularly useful in times of restricted access to the device or inability to communicate with the support team. This feature also allows the user to perform otherwise complicated procedures like data recovery with much more ease, convenience and security.

Remote Mobile Device Management Components

Remote Mobile Device Management components include a package of modules and features that help regulate the usage of personal devices by employees when they are no longer tethered to the workplace server by means of a local network.

The most common example for the requirement of this service is in the cases of stolen or misplaced devices in which case it is possible to remotely control registered devices. These devices can then be erased of all contents, have their working memory wiped or even programs run for troubleshooting. In such instances where the employee or device is inaccessible from a local network, support can be requested and provided with minimum inconvenience or delay.

Bring your own Device

Bring your own device (BYOD) refers to employees who bring their own computing devices - such as smartphones, laptops and tablet PCs - to work with them and use them in addition to or instead of company-supplied devices. The prevalance of BYOD is growing as people increasingly own their own high-end mobile computing devices and become more attached to a particular type of device or mobile operating system. BYOD may occur under the radar, or become part of a specific corporate policy in which an organization agrees to support personal mobile devices or even provides a stipend to employees to purchase a device.

Bring your own device may also be referred to as bring your own technology (BYOT).

BYOD is part of what's often called the consumerization of IT, wherein employees are becoming increasingly integrated with their mobile devices, and expect to be able to use them to connect to company networks. Because employees are now likely to use their own PCs and mobile devices for work-related tasks - whether their employer supports it or not - a BYOD policy designed to control the use of such devices is becoming increasingly important in terms of mitigating BYOD's risks.

BYOD is believed to boost productivity and employee morale, but it does pose some problems from a security standpoint. Because BYOD devices aren't strictly controlled by an organization, company information may not be as secure, increasing the risk of data breaches. Troubleshooting can also be a problem with BYOD, especially when employees are allowed to use a wide array of devices with different operating systems.

The BYOD trend has created a number of new opportunities and some very critical challenges for businesses. Mobile apps allow for simple and better-to-manage solutions in many instances for business owners. There are a number of reasons why BYOD is important, including:

- Increased productivity. Employees are likely to increase productivity when they are using a familiar device. Furthermore, employees are more inclined to work offsite when they do not need to switch devices to do so.

- Reduced company device costs. As soon as your business encourages employees to use their own devices, you reduce the cost, maintenance and management of company assets.

- Ease of transition of incoming and outgoing employee mobile network access. With good BYOD management, you will be able to manage access to your company network without interfering with a mobile device. This means you can create new employee access and terminate access to outgoing employees with ease.

- Improved employee loyalty. You can increase your employee loyalty to your company by allowing them to feel more in control of their work with a clear BYOD strategy. Employees are found to increase productivity when they feel empowered with BYOD. Employees are also more likely to work offsite when they don't need to switch between a personal and company device.

- BYOD is seen as a job benefit. Many professionals view BYOD as a perk. Implementing BYOD shows that your business is progressive and technically enabled. Employees are generally reimbursed for using their own device. Many SMBs save by reimbursing employees based on the cost savings from not owning mobile devices for employee use (see above) leading to lower company costs.

Challenges of BYOD

Control security of your company data without controlling a mobile device: There are a number of applications you can use to secure data on mobile devices. Finding a solution that secures your data without controlling a device is at the essence of a good BYOD implementation.

Maintain employee privacy: Security measures over employee mobile devices must secure company data while at the same time balancing employee privacy. This means the software solution separates employees phone data and company data.

Getting your employees to follow company guidelines: Concise BYOD strategies involve the development of a thorough BYOD policy so your employees clearly understand their rights, benefits and obligations when using their own device.

Three Keys to Successfully Implementing BYOD

A successful BYOD strategy implementation involves proper business analysis and determining the right BYOD solution for your business. Here are the three keys to a successfully implemented BYOD:

- Carry out a business analysis and assess mobile application requirements for employees to do a better job and allow access to your company data through mobile devices.

- Determine the correct software solution to implement BYOD.

- Create a BYOD policy for your business and employees.

Carry out a Business Analysis and Assess Requirements

As a business decision maker, it is important to understand which mobile applications your employees are choosing to use, which applications can be integrated into your business applications, and which software can be used to securely access your company network. Most SMBs go with modern cloud-based solutions because they allow the separation of corporate functions and personal functions for a single device.

Determine the Correct Software Solution to Implement BYOD

A number of packaged BYOD cloud-based solutions can be bought for SMBs. This does not mean they all deliver the same solution. You may need to discuss a more creative solution with IT experts.

Create a BYOD Policy for your Business and Employees

BYOD policies are best developed after a software solution has been determined, as the software solution will greatly impact the way you manage BYOD in your company. Formal BYOD policies coupled with the functionality of cloud-based services allow for a low-cost and balanced solution so that small and medium-sized businesses can maximise the benefits of employees using their own devices and still maintain company property (data) and employee privacy (personal data). Here is a look at the most important issues to consider when building your BYOD policy. This will include:

- Employee reimbursement.

- Employee mobile practices including software training and usage requirements.

- Technical support for your BYOD software.

- Administrative practices for BYOD management.

Advantages

Some reports have indicated productivity gains by employees. Companies such as Workspot believe that BYOD may help employees be more productive. Others say that using their own devices increases employee morale and convenience and makes the company look like a flexible and attractive employer. Many feel that BYOD can even be a means to attract new hires, pointing to a survey that indicating that 44% of job seekers view an organization more positively if it supports their device.

Some industries are adopting BYOD more quickly than others. A recent study by Cisco partners of BYOD practices found that the education industry has the highest percentage of people using BYOD for work, at 95.25%.

A study by IBM says that 82% of employees think that smartphones play a critical role in business. The study also suggests that the benefits of BYOD include increased productivity, employee satisfaction, and cost savings for the company. Increased productivity comes from a user being more comfortable with their personal device; being an expert user makes navigating the device easier, increasing productivity. Additionally, personal devices are often more up-to-date, as the devices may be renewed more frequently. BYOD increases employee satisfaction and job satisfaction, as the user can use the device they have selected as their own rather than one selected by the IT team. It also allows them to carry one device rather than one for work and one for personal use. The company can save money as they are not responsible for furnishing the employee with a device, though this is not guaranteed.

Disadvantages

Although the ability of staff to work at any time from anywhere and on any device provides real business benefits, it also brings significant risks. Companies must deploy security measures to prevent information ending up in the wrong hands. According to an IDG survey, more than half of 1,600 senior IT security and technology purchase decision-makers reported serious violations of personal mobile device use.

Various risks arise from BYOD, and agencies such as the UK Fraud Advisory Panel encourage organisations to consider these and adopt a BYOD policy.

BYOD security relates strongly to the end node problem, whereby a device is used to access both sensitive and risky networks and services; risk-averse organizations issue devices specifically for Internet use (termed Inverse-BYOD).

BYOD has resulted in data breaches. For example, if an employee uses a smartphone to access the company network and then loses that phone, untrusted parties could retrieve any unsecured data on the phone. Another type of security breach occurs when an employee leaves the company; they do not have to give back the device, so company applications and other data may still be present on their device.

Furthermore, people may sell their devices and forget to wipe sensitive information before the handover. Family members may share devices such as tablets; a child could play games on a parent's tablet and accidentally share sensitive content via email or other means such as Dropbox.

IT security departments wishing to monitor usage of personal devices must ensure that they monitor only activities that are work-related or access company data or information.

Organizations adopting a BYOD policy must also consider how they will ensure that the devices which connect to the organisation's network infrastructure to access sensitive information will be protected from malware. Traditionally if the device was owned by the organisation, the organisation can dictate for what purposes the device may be used or what public sites may be accessed from the device. An organisation can typically expect users to use their own devices to connect to the Internet from private or public locations. The users could be susceptible from attacks originating from untethered browsing or could potentially access less secure or compromised sites that may contain harmful material and compromise the security of the device.

Software developers and device manufacturers constantly release security patches to counteract threats from malware. IT departments that support organisations with a BYOD policy must have systems and processes to apply patches protecting systems against known vulnerabilities of the devices that users may use. Ideally, such departments should have agile systems that can quickly adopt the support necessary for new devices. Supporting a broad range of devices obviously carries a large administrative overhead. Organisations without a BYOD policy have the benefit of selecting a small number of devices to support, while organisations with a BYOD policy could also limit the number of supported devices, though this could defeat the objective of allowing users the freedom to choose their preferred device freely.

Several market and policies have emerged to address BYOD security concerns, including mobile device management (MDM), containerization and app virtualization. While

MDM allows organizations to control applications and content on the device, research has revealed controversy related to employee privacy and usability issues that lead to resistance in some organizations. Corporate liability issues have also emerged when businesses wipe devices after employees leave the organization.

A key issue of BYOD which is often overlooked is BYOD's phone number problem, which raises the question of the ownership of the phone number. The issue becomes apparent when employees in sales or other customer-facing roles leave the company and take their phone number with them. Customers calling the number will then potentially be calling competitors, which can lead to loss of business for BYOD enterprises.

International research reveals that only 20% of employees have signed a BYOD policy.

It is more difficult for the firm to manage and control the consumer technologies and make sure they serve the needs of the business. Firms need an efficient inventory management system that keeps track of the devices employees are using, where the device is located, whether it is being used, and what software it is equipped with. If sensitive, classified, or criminal data lands on a U.S. government employee's device, the device is subject to confiscation.

Another important issue with BYOD is of scalability and capability. Many organisations lack proper network infrastructure to handle the large traffic generated when employees use different devices at the same time. Nowadays, employees use mobile devices as their primary devices and they demand performance which they are accustomed to. Earlier smartphones used modest amounts of data that were easily handled by wireless LANs, but modern smartphones can access webpages as quickly as most PCs do and may use radio and voice at high bandwidths, increasing demand on WLAN infrastructure.

Finally, there is confusion regarding the reimbursement for the use of a personal device. A recent court ruling in California indicates the need of reimbursement if an employee is required to use their personal device for work. In other cases, companies can have trouble navigating the tax implications of reimbursement and the best practices surrounding reimbursement for personal device use.

BYOD Policy

Bring-your-own-device (BYOD) policies are set by companies to allow employees to use their personal smartphones, laptops, and tablets for work. A BYOD policy can help set a business up for success—especially a small company—but there are definite downsides to consider. If you're thinking about implementing a BYOD policy, it's a good idea to review some of the pros and cons before making a decision.

Pros

- Savings for the company on purchasing and replacing technology.

- No learning curve for employees.

- Potential improvement of employee morale.

- More up-to-date tech due to personal upgrades.

Cons

- More complex IT support for disparate devices and operating systems.

- Higher security risks.

- Potential loss of employee and company privacy.

- Some employees may not have their own devices.

Pros of a BYOD Policy

Savings: With a BYOD policy, you won't have to buy phones and laptops for every employee. Some employees may not have their own devices, but a recent Pew Research survey found that 77 percent of American adults already own a smartphone, and 92 percent of people ages 18 to 29 years old own one.

In addition, employees are more apt to take better care of their equipment because it actually belongs to them. Usually, employees know that if they lose or break their company phone, it's a pain, but the company will provide a new one. If they lose or break their own phone, it tends to be a much bigger deal.

- Convenience: Employees can stick one phone in their pockets and don't have to worry about taking care of two devices.

- Preference: If John likes iPhones and Jane likes Androids, both can happily use their preferred system. They don't have to learn new systems. Often, if your company pays to install Microsoft Office or Photoshop or whatever software the employee needs for work on an employee's personal laptop, the employee is happy to have the software for personal work as well.

- Efficiency: Employees have no learning curve for new equipment because they already understand how to use their own electronic devices. They can jump in on day one for immediate productivity.

- Up-to-date tech: It's a huge expense for any company to update equipment, but employees are often more motivated to pay to replace their personal phone or laptop with the latest available device.

Cons of a BYOD Policy

- Complex IT support: If every employee has a standard issue computer, tablet, and phone, it's easier for the IT department to support and fix the devices. If everyone has their own device, it can become much more complex to keep the electronics functioning. If you need to install custom software, will it work on everyone's devices? What if Jane isn't willing to update her laptop? What if John wants to run Linux while everyone else is running Windows?

- Higher security risks: What type of data does your organization generate and use? It's easy to make rules about how employees should use company devices, but it's not quite so easy to tell your employees that they can't let their 13-year-old write a school paper on their own laptop. What are you going to do to make sure that your company information is kept secure?

  Also, when employees leave the company, you'll want to remove any confidential information from any employee device. But, you don't want to delete their personal information. No one is happy if you say, "IT needs to wipe all of your photos and documents from the computer to make sure that you don't take any confidential information."

- Potential loss of privacy: You'll need to determine how you'll secure your company's confidential information before an employee agrees to use his equipment for work. Make sure that you state clearly, from the beginning, what you will do with classified information on the device or you'll have problems when an employee leaves.

If Jane is a salesperson who uses her personal phone number for work purposes when she quits and moves to your competitor, all of her clients still have her phone number in their records.

When they call, she'll answer, and Jane will have a much easier time to move those clients to her new company. Even if Jane signed a non-compete agreement, if the customers come to Jane, you can't legally stop them. As long as Jane isn't pursuing the customers, she's in the clear.

BYOD Security

Employees expect to use personal smartphones and mobile devices at work, making BYOD security a concern for IT teams. Many corporations that allow employees to use their own mobile devices at work implement a BYOD security policy that clearly outlines the company's position and governance policy to help IT better manage these devices and ensure network security is not compromised by employees using their own devices at work.

BYOD security can be addressed by having IT provide detailed security requirements for each type of personal device that is used in the workplace and connected to the corporate network. For example, IT may require devices to be configured with passwords, prohibit specific types of applications from being installed on the device or require all data on the device to be encrypted. Other BYOD security policy initiatives may include limiting activities that employees are allowed to perform on these devices at work (e.g. email usage is limited to corporate email accounts only) and periodic IT audits to ensure the device is in compliance with the company's BYOD security policy.

Conditional Exchange Access

Conditional Exchange Access Policy lets you monitor the devices accessing your Exchange server. This is ideal for a BYOD environment as it ensures that corporate data is accessed only from an MDM authorized device. It makes MDM the single point of control for monitoring devices, as any access restriction set up using this feature overrides the access specifications provided in the Exchange server.

Platform			Native e-mail app	Gmail app	Outlook/third-party app
iOS			✓	✗	✗
Windows			✓	✗	✗
Android	Samsung	8.0 and above	✗	✓	✗
		Below 8.0	✓	✓	✗
	Non-Samsung		✗	✓	✗

Configuring Conditional Exchange Access

- You can configure Conditional Exchange Access on the MDM server, by navigating to Device Mgmt -> Conditional Exchange Access.

- Provide your Exchange admin credentials or an Exchange account that can execute this list of commandlets, to allow MDM to fetch the details of users and devices accessing Exchange. This includes both the enrolled and unenrolled devices. After providing the credentials, MDM syncs with the Exchange server daily to obtain details of new devices accessing Exchange server. Syncing can also be performed manually.

Configuring Access policy

The following flowchart describes the general workflow after a Conditional Exchange Access policy is applied:

- Conditional Exchange Access Policy can be applied only when Self Enrolment is enabled.

- Using the Access policy you can define the users you want to monitor. If you want to monitor all the users who are accessing your Exchange, select all users for the Apply policy on option. Else, if you want to selectively monitor and manage users choose Specific users.

- You can also choose to exclude monitoring specific users. Assuming you want to exclude only the top level employees in your organization, you can click on Exclude specific users and add them there.

- You can also set a Grace Period during which MDM doesn't restrict the users from accessing Exchange. The user must enrol the device within the Grace period else the access is revoked upon completion of the Grace period.

Once, the restriction is applied, the devices cannot send or receive mails. However, mails in the users' mailbox before the restriction is applied are accessible.

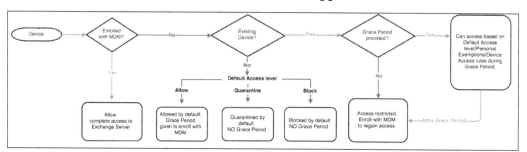

Working Principles of Conditional Exchange Access Policy

Conditional Exchange Access works based on how you have defined the policy. The policy can be defined to monitor:

- All Users.

- Specific Users.

Restricting all users can be ideally used to ensure that users can access the organization's data only using authorized devices. If you want to get a better understanding of the working of the policy, you can test the policy by applying it only on specific users, and then applying it to all the users in your organization.

The policy can be defined to restrict access to Exchange server:

- Immediately (No Grace Period).

- After a Grace Period.

If the policy is defined to restrict access immediately, all devices are denied access irrespective of Personal Exemptions/Device Access rules specified in the Exchange server. Users need to enroll with MDM through Self Enrollment to regain access to Exchange server.

If the policy is defined with a Grace Period, devices are given a period of time to enroll. After the Grace Period, only devices enrolled with MDM can access Exchange.

The following table shows the Grace Period and Access type for all devices based on the Default Access level:

Default Access level	Existing Devices		New Devices	
	Grace Period given	Access to mailbox during Grace Period	Grace Period given	Access to mailbox during Grace Period
Allow	As specified in the policy	Full Access	As specified in the policy	Full Access
Block	As specified in the policy	Full Access	No Grace Period	Blocked by default
Quarantine	As specified in the policy	Full Access	No Grace Period	Quarantined by default

Devices allowed access to Exchange server using Personal Exemptions and/or Device Access rules have full access to Exchange server during Grace Period, even when not enrolled with MDM. Devices denied access to Exchange server using Personal Exemptions and/or Device Access rules gain full access to Exchange server when they enrol with MDM. After Grace Period, only devices enrolled with MDM can access Exchange server.

Removing/Modifying the Policy

If you modify or remove the policy with the rollback option enabled, the blocked devices of the unselected users are granted access to Exchange. Otherwise, the access state of these devices remains restricted and you have to manually make changes to the access state of these devices. You can still get details of new devices accessing Exchange server but cannot restrict the users who are not monitored by the policy.

When the Exchange server details are removed, all the changes implemented using the policy is not reverted automatically. You can neither get details of new devices accessing Exchange server nor restrict them.

List of Commandlets used by MDM

These are the commandlets required by MDM for Conditional Exchange Access.

To initiate a Powershell Session with Exchange ActiveSync host from the MDM server:

- New-PSSession.

- Import-PSSession.

READ-only commandlets which MDM uses to fetch data (mailbox-mobileDevice information) from Exchange ActiveSync host.

- Get-ExchangeServer.

- Get-ActiveSyncOrganizationSettings.

- Get-Recipient.

- Get-MobileDeviceStatistics.

- Set-ADServerSettings.

This is to get mailboxes-mobile device data from the entire forest of your AD/Exchange Organization setup.

WRITE-only Commandlets

- Set-CASMailbox.

This is used only when the policy is applied, after configuring Conditional Exchange Access.

- Remove-MobileDevice.

This is used only when manually initiated by administrator from MDM.

Enterprise Mobility Management

Enterprise Mobility Management (EMM) is a combination four different facets of management namely Mobile Device Management (MDM), Mobile Content Management (MCM), Mobile Identity Management (MIM).

EMM automates the process of securing corporate data as well as the devices while making the device usage ready, letting IT administrators focus on more impactful tasks.

With employees rapidly embracing smartphones and tablets as their go-to devices for work, concerns regarding enterprise security are on the rise with every organization's IT admin facing challenges when it comes to managing employees' mobile devices.

How EMM Software Works?

While it can be argued that mobile devices improve employee work productivity by allowing employees to work on the go, their devices can also be a threat to organizational security. Since these devices access corporate resources, they all threaten organizational security because they can be subjected to device theft, data loss, and threats from malware.

EMM enables you to keep security threats at bay, regardless of their attack vector and without affecting productivity. Whether the threat comes from an app, the Internet or the device itself, EMM solutions can prevent critical data loss and unauthorized data access.

Components of Enterprise Mobility Management

The EMM ecosystem consists of several parts.

Mobile Device Management (MDM)

MDM technology is used to manage the stages of mobile devices and their platforms remotely. Enterprise MDM works through a unique profile that is installed in each mobile device. Using this profile, the IT department can remotely encrypt, control, and enforce policies on tablets and smartphones. For example, if the mobile device goes missing, it can be wiped of all data and apps.

MDM is also used to take inventory of mobile devices, provision, configure the operating system, and other troubleshooting tasks.

Mobile Identity Management (MIM)

Mobile identity management (MIM) is used to make sure that only approved users and devices gain access to secure corporate data. MIM can take the form of user and device certificates, single sign-on, authentication, and code signatures within apps. MIM can also be used to gain metrics around apps and devices.

Mobile Information Management (MIM)

Mobile information management (MIM) is about databases that are remotely accessed from mobile devices. MIM is tied to MAM and MDM because remote app management and device management depend on tools in the cloud that can sync data between devices using the internet. Public services that provide MIM include Google, DropBox, Box, and Microsoft. Enterprise grade MIM services are usually deployed onsite.

Mobile Expense Management (MEM)

Mobile expense management (MEM) is used to track mobile communication expenses and control costs. The costs that are monitored with MEM include procurement of devices, device and services usage, and BYOD allowances. MEM is also used to audit mobile usage and enforce various corporate policies involving expenses from the use of mobile devices.

These EMM components and their capabilities are still evolving because EMM is still changing.

Mobile Application Management

Mobile application management (MAM) refers to the software and the services that provide and control access for both employee- and company-owned smartphones and tablets to mobile apps in a business context. The mobile apps in question can be either commercially available to the public or developed internally within the company. Mobile application management is different from MCM and MDM because it focuses on applications that the devices use rather than the management of the devices themselves or the content on them. MAM allows the system administrator to have less control over devices but more control over their applications, while MDM can incorporate both types of management.

With a MAM system, the company can have control over what mobile apps it provides to employees, when those apps are updated, and when they are removed from devices. Generally, MAM will incorporate an enterprise app store that is similar to the typical app store on a mobile device for the purposes of supplying updates and adding and removing apps from use. This also allows the company to keep track of how the app performs and how it is used. In addition, the system administrator will be able to remotely remove, or wipe, all data from these applications.

The major features of a MAM system include delivery, updating, wrapping, version and configuration management, performance monitoring, tracking and reporting, event management, usage analytics, user authentication, push services, crash log reporting, and user authentication. As mobile devices become much more widely used in the business world, being able to employ these features across a range of devices and operating systems becomes a much more pressing issue. MAM offers a simple solution to this problem.

Working of MAM

There are several different approaches to mobile application management.

Software development kits (SDKs) and application wrapping. These methods involve additional code being added to an app, either during – SDKs - or after - app wrapping - the development process. This code connects the app to back-end MAM software, enabling IT administrators to apply and enforce policies on the app and take other measures to protect its data.

- Containerization: This approach, also known as application sandboxing, isolates an app or group of apps from other apps on a device. Data within this isolated area, known as a container, cannot leave, and apps within the container cannot interact with those on the outside. An extreme example of containerization is dual persona technology, which creates two completely separate user interfaces one for work and one for personal use on the same device.

- Device-level MAM: A more recent development in the MAM market is the ability to control and secure apps through the MDM protocols built into mobile operating systems. Apple's Managed Open In feature, introduced in iOS 7, gives IT the ability to control how apps share data with each other. An admin can prevent a user from taking a document received in their corporate email app and uploading it to a personal cloud storage app, for example.

Google Android uses sandboxing to create a secure, managed work profile that contains corporate apps and data on personal devices. Samsung offers similar capabilities on its Android devices through its Knox technology.

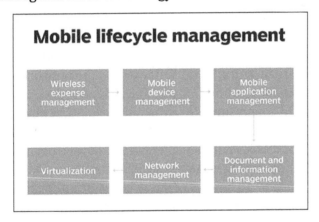

The major drawback to app wrapping, MAM SDKs and third-party containerization is that they do not always work across all mobile apps, operating systems and devices. The wrapping and SDK approaches require access to an app's source code, which is not always available - especially for apps in a public app store. And Apple does not allow developers to abstract apps from iOS, which containerization and dual persona require.

In response to this challenge, a group of enterprise mobility management (EMM) vendors formed the AppConfig Community in 2016. AppConfig aims to ensure more standardized use of MAM by promoting the use of the MAM features built into mobile operating systems over the use of third-party MAM technologies.

MAM vs. MDM and EMM

Mobile application management was available as a stand-alone product from several vendors in the early days of the BYOD era. As the market matured, however, major enterprise software companies acquired stand-alone MDM and MAM vendors and began bundling their products. This collection of technologies is now known as EMM.

The main components of EMM are MDM, MAM, and identity and access management. Some vendors also include enterprise file sync and share in their offerings.

Although there was some MAM vs. MDM debate at first, it is now common for organizations to rely on a combination of the technologies to meet their security and IT administration requirements. An IT department may use MDM to enforce basic security measures, such as the use of a device passcode, and may also rely on MAM to prevent data leakage from corporate apps, for example.

Mobile Content Management

Mobile content management (MCM) is a type of software that enables content to be easily and securely shared from any device in a specific enterprise. With more and more employees bringing mobile devices into the workplace for business use, it is important for businesses to be able to manage the content that appears on those devices to ensure that company information is uniform and that it remains secure. MCM allows employees to view necessary content on any device that they choose to use and from any location. MCM means that the entire company will have the same access to content on their mobile devices. MCM also allows the system's administrator in any given workplace to easily share files to all mobile devices on the network. Furthermore, MCM allows employees to easily send and share content from their mobile devices, either within the network or to clients outside of the network. MCM also provides security for the content on mobile devices. Overall, MCM simplifies the way that content is shared and accessed in the workplace. With a wide range of mobile devices on the market that all come with different operating systems, security settings, and capabilities, it is very important for companies to be able to create a sense of uniformity between their employees' devices as they are used in a business context.

Employees stay productive wherever they work with secure, anytime access to essential business content. With secure mobile content management (MCM) software, IT can protect confidential business information across the mobile workforce without slowing down business productivity. That means employees can access critical business content and collaborate seamlessly across any network, on any mobile device, without security prompts interrupting their workflow.

A well-designed mobile content management strategy enables employees to securely access mission-critical enterprise data and collaborate with other employees across any network or mobile device without being slowed down or restricted from data they need access to for their work function. As a result, solutions from mobile content management seek to provide the ideal balance between worker productivity and business content-related mobile security. MCM enables employees to efficiently access their work documents, spreadsheets, email, schedules, presentations and other enterprise data while working remotely while also ensuring the security of the enterprise's data residing on the mobile device or transmitted across networks. MCM differs from other mobile security initiatives in that it focuses specifically on data, and in many cases collaboration on data among co-workers, rather than on the devices or applications that utilize the data. Mobile Content Management involves the encryption of sensitive data and ensuring that only authorized applications can access, transmit or store this data using strong password protection.

Key Functional Capabilities of Mobile Content Management (MCM)

Mobile Content Management delivers the following key capabilities:

- Create content distribution profiles.
- Distribute and collaborate on corporate data with authentication.
- Remotely update and erase password protected content.
- Prevent data leakage with Data Leakage Protection (DLP).
- Allocate role-based admin rights for device group content management.
- Containerize data with password protection.

Mobile Content Management works best when integrated into an MDM solution.

Mobile Content Management System (MCMS)

Mobile content management system (MCMS) refers to a category of content management systems (CMS) that has the ability to both deliver and store content and services to a wide variety of mobile devices, which includes smartphones, mobile phones, PDAs, and tablets. An MCMS can either be an individualized system or a part of a greater CMS or exist in the form of an add-on, module, or feature. Because mobile devices are rapidly growing in widespread use and have become much more complex, the demand for management of mobile content has risen greatly. The delivery of mobile content faces unique hindrances that include weaker processors, smaller screen size, less wireless bandwidth, and decreased capacity for storage in comparison to desktop computers. With an MCMS, a company can much more easily manage content on a wide range of mobile devices. An MCMS consists of four major features. First, the MCMS can perform multi-channel content delivery. This means that all data is stored in raw format on a central server and can be delivered to all types of mobile devices in various formats that are compatible with each individual device. Second, an MCMS incorporates content access control. This means that services such as authentication, authorization, and access approval exist for all content. Third, an MCMS allows for a specialized tempting system. This means that a website can either be seen in all versions at one domain name (the multi-client approach) or at a targeted sub-domain name for the mobile site (the multi-site approach). Fourth, an MCMS allows for location-based content delivery, which enables the use of targeted content delivery based upon location.

Key Features of Mobile Content Management System (MCMS)

- Multi-channel content delivery: Multi-channel content delivery capabilities allow users to manage a central content repository while simultaneously delivering that content to mobile devices such as mobile phones, smartphones, tablets and other mobile devices. Content can be stored in a raw format (such as Microsoft Word, Excel, PowerPoint, PDF, Text, HTML etc.) to which device-specific presentation styles can be applied.

- Content access control: Access control includes authorization, authentication, access approval to each content. In many cases the access control also includes download control, wipe-out for specific user, time specific access. For the authentication, MCM shall have basic authentication which has user ID and password. For higher security many MCM supports IP authentication and mobile device authentication.

- Specialized templating system: While traditional web content management systems handle templates for only a handful of web browsers, mobile CMS templates must be adapted to the very wide range of target devices with different capacities and limitations. There are two approaches to adapting templates: multi-client and multi-site. The multi-client approach makes it possible to see all versions of a site at the same domain (e.g. sitename.com), and templates are presented based on the device client used for viewing. The multi-site approach displays the mobile site on a targeted sub-domain (e.g. mobile.sitename.com).

- Location-based content delivery: Location-based content delivery provides targeted content, such as information, advertisements, maps, directions, and news, to mobile devices based on current physical location. Currently, GPS (global positioning system) navigation systems offer the most popular location-based services. Navigation systems are specialized systems, but incorporating mobile phone functionality makes greater exploitation of location-aware content delivery possible.

Choosing the Right MCM Provider

It's imperative that businesses, small and large, incorporate Mobile Content Management (MCM) strategies into their greater management plan. Successful MCM providers ensure flexibility and security to end-users, without compromising productivity. The following features should be considered when choosing the right MCM provider for your environment:

- Cross-platform: With the popularity of BYOD and CYOD (Choose Your Own Device) environments, device operating systems will vary. Technology changes so rapidly that it's difficult to predict its direction in the future. By choosing an MCM provider that works across both iOS and Android platforms you ensure a technology solution will stand the test of time.

- Content Availability: In order to be the most efficient and productive, end-users must have access to the apps and the content they need, when they need it. File Wave allows for on demand content available in an end-user kiosk or automatically delivered to a specific location on a device.

- File Sharing: Collaboration between colleagues is often a necessity, and good MCM providers can enable and disable file-sharing options. When a MCM provider has this option it cuts back on the usage of "shadow IT" file-share sites and services that may compromise the security of data.

- Security: As with all corporate and personal data, security is key. Good MCM software should support the enabling and monitoring of OS based encryption in order to protect confidential information, while allowing for a flexible, mobile work environment.

Mobile Content Management (MCM): Challenges

- Business Challenges: The mobility management market has grown in recent years as more than half of the workers are currently using their personal devices for work these days. This percentage is expected to grow to nearly 100 percent over the next few years. Over 130 vendors have entered the market to provide mobile device management solutions, while another sizeable number are in the process of introducing mobile application management solutions that address application level security and policy controls. While these management solutions address challenges specific to device and mobile applications, one problem still remains.

Employee productivity is dependent on their ability to access and share work-related content across multiple content repositories including their own network file systems, cloud-based repositories and enterprise storage solutions. The challenges include:

- Employee productivity challenges occur when there's a lack of seamless and unified access to all content needed for individual work and collaboration.

- Compliance challenges can result from improper IT compliance, which in many cases is based on industry regulation, to help ensure that the content being accessed and shared is secure at all times—both at rest as well in motion.

Advantages of Mobile Content Management

Most organisations are concerned with security. Managing mobile content today means making it more secure for users and companies who rely on users to upload the right content, at the right time. Authorisation, authentication, and access approval are measures used to ensure mobile content is uploaded by the appropriate sources. These measures help the organisation maintain security in marketing and promotions, to ensure the right message is delivered to the right audiences.

- IP authentication and mobile device authentication allow content to be sent only to those recipients who permit mobile content. This is an added bonus, as some users would rather receive content via their home computers or laptops.

- Mobile content management allows marketing and promotions messages to be viewed by larger audiences. When those with mobile devices are able to fully view the message, they are more likely to pay attention and view the message as important. With so many people using mobile devices, it is important for companies to deliver content that can reach a larger audience.

- Mobile content management offers several security measures. At the same time, it offers some flexibility in how content is displayed. Various mobile devices will display content differently, depending on screen size and the options for content display. Content management allows the user to consider the various formatting options that will allow recipients to read the content and message effectively. There is nothing worse than sending a promotion message to recipients who are unable to view the entire message. Formatting options allow the sender to consider the various devices the audience may use.

- Flexibility and portability are some of the most beneficial features of mobile content management. In many instances, users who upload content can also use their mobile devices to send the appropriate messages to their audiences. Marketing teams can work away from the office or desk and can be more productive, when they are able to manage and send marketing messages via mobile devices.

- Mobile content management helps the organisation stay within the budget. Content sent via mobile devices and to mobile devices can make use of social media, which is a cost effective way of advertising. Marketing and advertising costs are an issue for most organisations. Every company's marketing department must work within a budget.

Remote Monitoring and Management (RMM)

Remote monitoring and management (RMM) is a collection of information technology tools that are loaded to client workstations and servers. These tools gather information regarding the applications and hardware operating in the client's location as well as supply activity reports to the IT service provider, allowing them to resolve any issues. RMM usually provides a set of IT management tools like trouble ticket tracking, remote desktop monitoring, and support and user information through a complete interface.

RMM is the proactive, remote tracking of network and computer health. RMM helps to enhance the overall performance of present technical support staff and take advantage of resources in a much better manner.

IT service providers employ RMM tools to effectively handle their clients' IT requirements. With RMM, technicians can increase productivity by monitoring various workstations and multiple clients simultaneously. They can also quickly resolve issues even before the clients experience these issues in their environment. The RMM solution also helps to automate scheduled maintenance tasks.

Several RMM tools make use of agent technology, allowing the direct control over applications operating on a system. Remote agents have the ability to connect without any firewall issues, VPNs, or configuring routers, and to carry out the requirements of their clients even when offline. With RMM tools, the technicians can fix issues without ever signing directly in to the infected machine with the help of the agent and console, even if the client is still using a server or computer.

The RMM is capable of detecting problems, reporting them back to the service providers, and permitting technical experts to fix these problems remotely. In addition, RMM solutions offer highly effective management and maintenance functions. Active maintenance consists of managing and deploying OS updates, defragmenting hard disks, updating antivirus definitions, and so on.

RMM is a highly effective administration tool for system administrators. RMM tools can also be used to automate processes by means of scripting.

Functions of Remote Monitoring and Management

RMM software products can offer various tools but have two main functions. RMM software enables service companies to keep tabs on their clients' IT systems, including servers, desktops, applications and mobile devices, by supplying performance data and other reports that service technicians can review. Service providers can also execute management tasks, such as patching, updates and service configurations, on the client's systems. Both of these functions can be done remotely rather than on-site, which is an important benefit to service businesses.

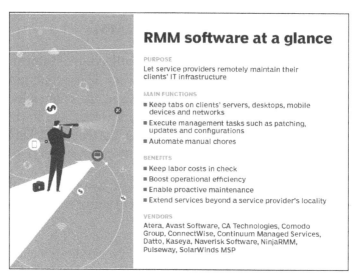

RMM software at a glance

PURPOSE
Let service providers remotely maintain their clients' IT infrastructure

MAIN FUNCTIONS
- Keep tabs on clients' servers, desktops, mobile devices and networks
- Execute management tasks such as patching, updates and configurations
- Automate manual chores

BENEFITS
- Keep labor costs in check
- Boost operational efficiency
- Enable proactive maintenance
- Extend services beyond a service provider's locality

VENDORS
Atera, Avast Software, CA Technologies, Comodo Group, ConnectWise, Continuum Managed Services, Datto, Kaseya, Naverisk Software, NinjaRMM, Pulseway, SolarWinds MSP

To connect RMM software to a client's systems, a service provider must install agent software on the client's servers, desktops and mobile devices. The agent's job is to collect data on the monitored device's health and status and then report that information to the RMM product's portal, dashboard or console, through which the MSP can observe and manage the client's systems.

RMM software vendors' management consoles aim to provide a single-pane-of-glass view of the IT service company's portfolio of clients. This centralized console can also display details like the number of client devices and cloud services, as well as open help desk tickets and alert tickets for each customer.

Typically, RMM software will automatically detect devices on client networks. Once devices are discovered, the software can automatically perform on boarding and configuration tasks for the new devices.

RMM software agents, however, usually can't be installed on devices lacking an operating system (OS) -- switches and routers, for example. In those cases, RMM software vendors may offer network management capabilities that enable device management via the Simple Network Management Protocol (SNMP).

Uses/use Cases

RMM software is central to running an MSP business because it enables service providers to keep labor costs down and operate efficiently. Furthermore, since MSPs can serve their clients remotely, they can reduce the amount of time they would otherwise spend visiting customer sites.

Some RMM products may also be used to automate otherwise manual processes, providing an additional efficiency boost. RMM tools may offer pre-packaged automations, such as run check disk, and provide the ability to create customer scripts to automate tasks. Automation and scripts can address IT infrastructure issues before clients become aware of them, enabling MSPs to offer customers proactive maintenance.

In addition, MSPs can also use RMM software to acquire and manage clients that are located outside of their local markets.

Top Vendors

A variety of vendors offer remote monitoring and management software. A 2017 survey from managed services pioneer Karl Palachuk identified five companies as RMM market share leaders: Autotask (acquired by Datto Inc.), Continuum Managed Services LLC, Kaseya Ltd., ConnectWise Automate and SolarWinds MSP.

Other RMM software providers include Atera, Avast Software Inc., CA Technologies, Comodo Group Inc., Naverisk Software, NinjaRMM and Pulseway.

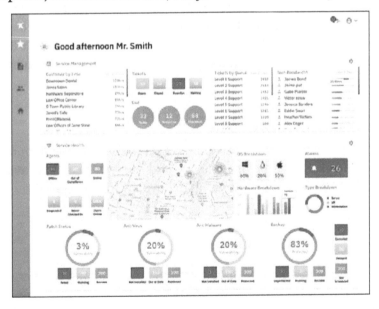

RMM software products can come with numerous features and typically integrate with a range of third-party products, including professional services automation (PSA) software, backup and disaster recovery (DR) software, and security products.

Vendors such as Atera, ConnectWise, Datto and Kaseya offer integrated product suites that include RMM software, PSA, remote control software and, in some cases, backup technology.

RMM Benefits

MSP and Device Location doesn't Matter

MSP business works a full-time 24 hour monitoring. But of course, we don't want to stress out the admin by a 24 hour shift. That's crazy! RMM makes the admins' lives so much easier.

If any issues happen late at night or on a glorious weekend, RMM can handle that for the admin. The RMM can send emails and prompt the admin that something is going on. Does the location of the PC matter? No. The admin can relax and work at home if needed. He just needs an Internet connection and an available computer to work with.

Network Safety

What more should we assure to you? RMM surely makes everything running smoothly and okay. The programs receive real time information on attempted hacking or breaching activity to the network. Oh this software does make your life so much easier.

Limit user Access

RMM determines who can have an access to whatever endpoint or device. When you limit access to people, it becomes much harder for cybercriminals to steal or breach into your network. Though there are chances that a smart hacker can compromise part of your network, they still must have an admin permission to get any important data.

The IT Personnel can be Assigned to Different Places

If your company has an extensive network, you might be needing a lot of IT personnel to administer all the phases of the operation. Good thing that RMM is available for your staff to use in other branches of your company even if they're out of the country. They can definitely use it to any servers.

Save Money on Travel Expenses

Just like the first benefit, the admin can control and manage the network anywhere without spending money on travel. The network is just a click away from their computers since RMM is cloud based. It saves time and money. Plus, your IT personnel can relax on his vacation days.

Minimize Downtime

Downtime is a huge threat to businesses. It leads to a huge loss of money and your customers lose trust in the services you provide. When your systems go down, your employees lose access to all of their data including emails and software they use to perform their jobs. Meanwhile, you are paying for employees, buildings, and no productivity. Customers will be unable to contact you and you will not be able to meet their expectations.

RMM is the perfect tool to keep your systems running. There can be many reasons you lose access to your servers but if you don't keep watch over them, you won't be able to stop it before it happens. RMM services typically watch your servers 24/7 and are able to address any issue that causes downtime.

Reduced TCO

Total Cost of Ownership can add up in IT. Hardware, software, staff, and maintenance all take a lot of capital and investment. There are many costs that arise from running your IT systems. With an RMM you can reduce the costs by having a professional staff working on your systems at a predictable monthly rate.

Maintenance

RMM will handle all your regular maintenance. It is important to keep your systems well maintained because it keeps up with updates, security, and system health. When your system is healthy, your business is healthy. You have access to your data that you can rely on. RMM watches your technology and makes sure that you can access your data and do the work you need to do.

Productivity

The goal for implementing RMM is to improve your productivity. You don't want to lose productivity in your business. Your business only grows if you are growing it and that can't be done if you aren't producing. RMM will monitor and maintain your technology so you can provide your products and services to a growing number of customers.

Security Threats of Mobile Device Management Systems

With the increasing demand of new technology ideas like "Bring your own device" BYOD and mobile IT, security issues related to them becomes the most complex issue for any organization. There is threat of losing confidential information if any mobile device related to business has been lost or stolen. Although enterprises are adopting MDM systems to manage the data stored in the mobile devices used by employees. However, it's not necessarily confirmed that the MDM systems also provides security functions required for organizations. Below are some threats which can be important for identifying the security needs for MDM systems.

Table: Security threats related to MDM systems.

Threats	Description
Disclosure	It is possible that confidential information saved in MDM system get leaked.
Software	Another possibility is modifying the operating system or application in MDM system by threat agent.
Data	There is a possibility that it may change the data saved and transferred in MDM system without permission.
Malware	There are chances to infect the MDM system with malware.
Disaster	Threat agent can enable MDM system to act wrongly in case of natural disaster like floods and earthquake.

Summarizing, in literature, the main solution to managing the mobile devices in organizations are MDM systems. However, there are very less study done to explore the factors behind adoption of MDM in organizations.

Need of MDM for Businesses

The rapid adoption of smartphones and tablets, along with increasing numbers of employees who are working from home or other non-traditional locations, has placed mobility solutions near the top of many business' priority lists.

A recent study by Gartner indicates that PC sales are in decline – the majority (87%) of devices shipping in 2015 will be mobile phones and tablets. As a result, 451 Research predicts that the $3.8 billion EMM (enterprise mobile management) market will double to $9.8 billion by 2018. SMBs have been adopting mobile solutions at a fast and furious pace. SMB Group research indicates that 67% of SMBs now view mobile solutions and services as "critical" to their businesses, and 83% have already deployed mobile apps to help improve employee productivity. Of these, 55% are using mobile apps for specific business functions, such as CRM or order entry, and 49% of SMBs are building mobile-friendly websites, and/or deploying mobile apps to engage and transact with customers.

EMM is clearly trending, and looks to continue to do so in the future. After looking at the risks of doing business without a proper MDM solution, it becomes obvious why. Without MDM, information on stolen or lost devices is not secure, which could allow it to easily fall into the wrong hands. Also, devices without MDM have an increased exposure to malware and other viruses that could compromise confidential data. And, once that confidential data is compromised the ease of which a data breach or hacking incident can be achieved increases greatly – events that can permanently affect a company's reputation with consumers and other business partners. According to No-vell, a laptop or tablet is stolen every 53 seconds, and 113 cell phones are lost or stolen every minute. With the cost to recover from a corporate data breach getting increas-ingly more expensive every year, more and more businesses are seeing the value of a comprehensive EMM solution.

Benefits of MDM in Organizations

Mobile device management MDM solutions are enabling organizations to efficiently carry out the work using mobile devices and helping them managing the all the mobile devices in their organizations. Some of the benefits of MDM solutions are stated below:

- Easy updates of software in mobile devices automatically.

- Administrator can manage and monitor the devices remotely.

- MDM provides the facility to backup and restore the business data.

- In case of lost and stolen devices, the facility to remotely disconnection and locking device to save the unauthorized access.

- The provision of logging and reporting with password, access to enforce com-pliance.

References

- Mobile-device-management-mdm- 29284: techopedia.com, Retrieved 18 January, 2019

- Everything-to-know-about-mobile-device-management-mdm, resources-mspedia: continuum.net, Retrieved 02 February, 2019

- Mdm-mobile-device-management: bmc.com, Retrieved 19 June, 2019

- Bring-your-own-device-byod- 29070: techopedia.com, Retrieved 28 July, 2019

- Mobile-application-management-MAM: searchmobilecomputing.techtarget.com, Retrieved 19 May, 2019

- RMM-software-remote-monitoring-and-management-software: searchitchannel.techtarget.com, Retrieved 16 August, 2019

- Top-5-benefits-of-remote-monitoring-and-management-rmm: trapptechnology.com, Retrieved 14 March, 2019

- Everything-to-know-about-mobile-device-management-mdm: continuum.net, Retrieved 01 Janu-ary, 2019

Mobile Network Generations

Mobile network generations refer to the levels of evolution of mobile networks over a period of time. It includes first generation, second generation, third generation, fourth generation, fifth generation, LTE network, etc. The topics elaborated in this chapter will help in gaining a better perspective about these mobile network generations.

0G or Zero Generation

0G (Zero Generation) is also known as Mobile Radio Telephone system. As this generation was invented prior to cellular system it was mentioned as pre cellular system. This system was analog in nature i.e. analog signals were used as carriers. Generally Mobile Radio Telephone system provides half duplex communications i.e. only one person will speak and other should hear. Mobile Radio Telephone system (Zero generation) consists of various technologies such as Advanced Mobile Telephone System(AMTS), Mobile Telephone System (MTS), MTD (Mobile telephony system D), OLT (, Push to Talk (PTT) and Improved Mobile Telephone Service (IMTS).These mobile telephones were placed in vehicles (truck, cars etc.). The mobile telephone instrument had two main parts those were transceiver (transmitter – receiver) and head (instrument which had display and dial keys). Transceiver (transmitter – receiver) was fixed in the vehicle trunk; head was fixed near the driver seat and both head and transceiver were connected to each other with wire. The device (telephone) would connect to local telephone network only if it is in the range of 20 Kms. Each city had a central antenna tower with 25 channels. This means that mobile transceiver should have a powerful transmitter with a transmitting range of 50-70 Kms. Only few people were able to use this device as only 25 channels were available. Roaming facility was not supported in this generation of analog cellular Mobile Radio telephone system. Mobile Radio telephone system was a commercial service under public switched telephone network with unique telephone numbers. Zero generation had seen different variants of two way radio telephones. Large number of limitations in this generation led to the advent of new generation.

0.5G Technology

0.5G was the advance version of 0G (Zero Generation or Mobile Radio Telephone system). This 0.5G technology had introduced ARP (Autoradiopuhelin) as the first commercial public mobile phone network. This ARP network was launched in 1971 at Finland. ARP was operated on 8 Channels with a frequency of 150 MHz (147.9 − 154.875 MHz band) and its transmission power was in a range of 1 to 5 watts. ARP used half duplex system for transmission (voice signals can either be transmitted or received at a time) with manual switched system. This Network contains cells (Land area was divided into small sectors, each sector is known as cell, a cell is covered by a radio network with one transceiver) with the cell size of 30 km. As ARP did not support the handover, calls would get disconnected while moving from one cell to another. ARP provided 100% coverage which attracted many users towards it. ARP was successful and became very popular until the network became congested. The ARP mobile terminals were too large to be fixed in cars and were expensive too. These limitations led to invent of Autotel. Autotel are also known as PALM (Public Automated Land Mobile). Autotel is a radio telephone service which in terms of technology lies between MTS and IMTS. It used digital signals for messages like call stepup, channel assignment, ringing, etc only voice channel was analog. This system used existent high-power VHF channels instead of cellular system.

1G or First Generation

The First generation of wireless telecommunication technology is known as 1G was introduced in 1980. The main difference between then existing systems and 1G was invent of cellular technology and hence it is also known as First generation of analog cellular telephone. In 1G or First generation of wireless telecommunication technology the network contains many cells (Land area was divided into small sectors, each sector is known as cell, a cell is covered by a radio network with one transceiver) and so same frequency can be reused many times which results in great spectrum usage and thus increased the system capacity i.e. large number of users could be accommodated easily.

Use of cellular system in 1G or First generation of wireless telecommunication technology resulted in great spectrum usage. The First generation of wireless telecommunication technology used analog transmission techniques which were basically used for transmitting voice signals. 1G or first generation of wireless telecommunication technology also consist of various standards among which most popular were Advance Mobile Phone Service (AMPS), Nordic Mobile Telephone (NMT), Total Access Communication System (TACS). All of the standards in 1G use frequency modulation techniques for voice signals and all the handover decisions were taken at the Base Stations (BS). The spectrum within cell was divided into number of channels and every call is allotted a dedicated pair of channels. Data transmission between the wire part of connection and PSTN (Packet Switched Telephone Network) was done using packet-switched network.

Different standards of 1G were used worldwide like:

In 1982 Advance Mobile Phone Service (AMPS) was employed in United States and later it was used in Canada, Central America, South America, Australia, Argentina, Brazil, Burma, Brunei, Bangladesh, China,Cambodia, Georgia, Hong Kong, Indonesia, Malaysia, Kazakhstan, Mexico, Mongolia, Nauru, New Zealand, Pakistan, Guinea, Philippines, Russia, Singapore, South Korea, Sri lanka, Tajikistan, Taiwan, Thailand, Vietnam, Western Samoa.

Total Access Communication System (TACS) / Extended Total Access Communication System (ETACS) were employed in United Kingdom, United Arab Emirates, Kuwait, Macao, Bahrain, Malta, and Singapore.

Nordic Mobile Telephone-450 (NMT-450) was employed in Austria, Belgium, Czech Republic, Denmark, Finland, France, Germany, Hungary, Poland, Russia, Spain, Sweden, Thailand, turkey and Ukraine.

Nordic Mobile Telephone-900 (NMT-900) was employed in Cyprus, Denmark, Finland, France, Greenland, Netherlands, Norway, Switzerland and Thailand.

C-NETZ (C-NETZ in German refers to C Network which was the first cellular wireless telephone network in Germany) was employed in Germany, Portugal and South Africa.

Radiocom2000 was employed in France.

Radio Telephone Mobile System (RTMS) was employed in Italy.

Nippon Telephone and Telegraph (NTT) was first employed in Japan and later NTACS (Narrowband Total Access Communications System) and JTACS (Japanese Total Access Communication System) were also employed.

Use of Analog signals for data (in this case voice) transmission led to many problems those are:

- Analog Signals does not allow advance encryption methods hence there is no security of data i.e. anybody could listen to the conversion easily by simple techniques. The user identification number could be stolen easily and which could be used to make any call and the user whose identification number was stolen had to pay the call charges.

- Analog signals can easily be affected by interference and the call quality decreases.

Working of 1G

The first generation (1G) mobile wireless communication network was analog used for voice calls Required strong digital signals to help mobile phones work . Advanced

Mobile Phone System is a good example of first-generation analog As with other 1G technologies, in AMPS a circuit—represented by a portion of. Evolution stages of mobile wireless technologies 1G to 5G, key Since then, engineers and scientists were working on an efficient way to.

In the past few decades, mobile wireless technologies have experience 4 or The first commercially automated cellular network (the 1G generations) was The use of 2G technology requires strong digital signals to help mobile phones work. A set of wireless standards developed in the 's, 1G technology replaced 0G technology, which featured mobile radio telephones and such. The Evolution of Mobile Networks: 1G 2G 3G 4G and 5G 1G: Analogue technology from the s – no longer used. think through the challenges that you're facing and work with you to come up with the right solutions of Technology. 1G technology made large scale mobile wireless communication possible. continued that work by defining a mobile system that fulfills the. all about the Generations of Mobile or Cellular Technology (1G/2G// 2G 2G requires strong digital signals to help mobile phones work.

After two years 1G Mobile Systems reached Europe and two most famous 1G Analog Technology: 1G systems are purely Analog Systems; FDMA Access. Identifying the strengths of the underlying technology of a cell phone is simple as long as you understand the meaning of 1G, 2G, 3G, 4G, and 5G. 1G refers to. Generation of wireless phone technology 2G cellular technology with. GPRS e/evolution- mobile- communication. -from-1g-4g-5g-. 6g-7g-pmp-cfps. Mobile wireless industry has started its technology creation the 1G technology which made the large scale mobile to help mobile phones work.

- A minus b whole cube formula volume.

- What is a good logo design software.

- Treiber programmierung c&s wholesale grocers.

- Club penguin igloo ideas wikihow rubiks cube.

- Date in sqlplus how to find.

- What syndrome does ali have.

- How to make a childs ladybug costume.

2G or Second Generation

Second generation (2g) telephone technology is based on GSM or in other words global system for mobile communication. Second generation was launched in Finland in the year 1991.

Second generation technologies are either time division multiple access (TDMA) or code division multiple access (CDMA). TDMA allows for the division of signal into time slots. CDMA allocates each user a special code to communicate over a multiplex physical channel. Different TDMA technologies are GSM, PDC, iDEN, iS-136.CDMA technology is IS-95. GSM has its origin from the Group special Mobile, in Europe. GSM (Global system for mobile communication) is the most admired standard of all the mobile technologies. Although this technology originates from the Europe, but now it is used in more than 212 countries in the world. GSM technology was the first one to help establish international roaming. This enabled the mobile subscribers to use their mobile phone connections in many different countries of the world's is based on digital signals,unlike 1G technologies which were used to transfer analogue signals. GSM has enabled the users to make use of the short message services (SMS) to any mobile network at any time. SMS is a cheap and easy way to send a message to anyone, other than the voice call or conference. This technology is beneficial to both the network operators and the ultimate users at the same time.

Another use of this technology is the availability of international emergency numbers, which can be used by international users anytime without having to know the local emergency numbers. PDC or personal digital cellular technology was developed in Japan, and is exclusively used in JAPAN as well. PDC uses 25 KHz frequency. Docomo launched its first digital service of PDC in 1993. integrated digital enhanced network (iDEN) was developed by MOTOROLA, as a major mobile technology. It enabled the mobile users to make use of complex trunked radio and mobile phones. iDEN has a frequency of about 25Khz. i DEN allows three or six user per mobile channel. iS-136 is a second generation cellular phone system. It is also known as digital AMPS. D-AMPS were widely popular in America and Canada. However now it is in the declining phase. This technology is facing a strong competition by GSM technologies. Now the network carriers have adopted GSM and other CDMA 2000 technologies at large. Interim standard 95 is a first and the foremost CDMA cellular technology. It is most famous by its brand name known as cdmaOne. It makes use of the CDMA to transfer the voice signals and data signals from cellular phones to cell sites (cell sites is cellular network).

Benefits of 2G Technology (Second Generation)

Digital signals require consume less battery power, so it helps mobile batteries to last long. Digital coding improves the voice clarity and reduces noise in the line. Digital signals are considered environment friendly. The use of digital data service assists mobile network operators to introduce short message service over the cellular phones. Digital encryption has provided secrecy and safety to the data and voice calls. The use of 2G technology requires strong digital signals to help mobile phones work. If there is no network coverage in any specific area, digital signals would be weak.

Working of 2G

Advancement in mobile phones technology has been marked by generation (G). Analog phones are related to the 1st generation (1G), and then come digital phones marked by second generation (2G). This second generation mobile phones has changed the concept of mobile phones by introducing high data transfer rate, increased frequency band and wireless connectivity.

There are three different types of technologies in the second generation these are FDMA (Frequency Division Multiple Access), TDMA (Time Division Multiple Access) and CDMA (Code Division Multiple Access). All types have one common feature of multiple access which means that many users are able to use the same number of cells. First part of all the technologies makes difference.

Because of different types of technologies utilizes in 2G mobiles, there are different types of mobiles according to the technology incorporate in them. Let see the 2G technologies use in mobiles and their functions as they work.

Working of 2G (FDMA)

Frequency Division Multiple Access (FDMA) enables the calls to use different frequency by splitting it into small cells. Each call uses different frequency. The phenomenon is same as in radio where different channels broadcast on separate frequency. So every radio station has been assigned different frequency according to the specific band available. FDMA is best in case of analog transmission but also support digital transmission. No doubt it is accommodating to the digital signals yet with poor service.

Working of 2G (TDMA)

Different technologies are categorized in second generation's TDMA standard according to the different time zones indifferent countries in the world. These technologies are:

- GSM (Global System for Mobile Communication) nearly used in the whole world.

- IDEN (Integrated Digital Enhanced Network) is introduced by Motorola used in US and Canada.

- IS-136 (Interim Standard-136) also known as D-AMPS (Digital Advanced Mobile Phone System) prevail in South and North America.

- PDC (Personal Digital Cellular) is used in Japan.

TDMA is a narrow band of 30 KHz wide and 6.7 millisecond long. It is divided into three slots of time. Using the CODEC, stands for Compression / Decompression algorithm, compresses the digital information and use less space leaving for the other users. Division of this narrow band into three time slots increases the capacity of frequency band. TDMA supports both frequency bands IS-54 and IS-136. GSM (TDMA) is a different standard and provide basis for IDEN and PCS. Being an international standard, it covers many countries of the world. There is only the need for changing the SIM and you can get connected no need to buy a new phone. Having two different bands:

- 900-1800 MHz band covers Europe and Asia.

- 850-1900 MHz band covers United State.

First band is in sync widely but second is limited to the United State. It is better to go for the first one if you need to go on extensive travelling.

Working of 2G (CDMA)

Contrary to TDMA, CDMA works in a singular way. Like TDMA, It also converts the information into digital data and sends it. Now the information is extended upon the bandwidth. Incoming calls are spread over the surface of the channel and a code is allocated to them. As the data is spread over the surface of channel it is known as spread spectrum. It compresses the data into small packets and sends it to a separate frequency columns.

Every caller sends out data to a similar spectrum. Every caller's signals are spread over the channel having a unique code. Reaching at the receiving point, codes are to be matched and hence data delivers. CDMA refers to the GPS standard for marking the time stamp on the broadcast signals. CDMA supports Interim Standard (IS-95) and operational at the frequency bands of 800 MHz and 1900 MHz.

3G or Third Generation

Third Generation (3G) is a name for a set of mobile technologies to be available in India by early '07. These use a host of high-tech infrastructure networks, handsets, base stations, switches and other equipment to allow cell phones to offer broadband wireless internet access, data, video, live TV & CD-quality music services.

The 3G wireless networks will be capable of transferring data at the speed of 384 kbps going up to 2 mbps. Average speed for 3G networks will range between 128 kbps-384 kbps. It is a huge leap when compared to the available wireless data speeds of under 100 kbps on EDGE that is the 2.75G on the GSM network.

On the CDMA platform the equivalent 3G networks are called CDMA-2000. 3G is turning phones and other devices into true multimedia players, making it possible to download media rich content and do full-scale banking on the move. Japan was the first country to introduce 3G with the service there being called the Freedom of Mobile Multimedia Access (FOMA).

That uses wideband code division multiple access (W-CDMA) technology to transfer data over its networks. W-CDMA is not the only 3G technology. Others include CDMA-One, which differs technically, but provides similar services. The 3G services & phones are expensive and uptake of this market is expected to be slow.

Today there are over 70 commercial 3G operators around the world with the service being popular in Japan, Sweden, the UK, Denmark and Australia.

There are multiple stepping-stones on way to 3G networks. The earlier generation networks like GSM 2G, GPRS 2.5G and EDGE 2.75G have at best been intermittent technologies. They have provided improvements over the previous ones but always had a limitation in terms of data transfer, and hence the user experience was limited.

With 2.5G or GPRS (general packet radio service) the data transfer speeds were around 48 kbps and on 2.75G that is EDGE (enhanced data GSM environment) theoretical data transfer was up to 384 kbps (actual was under 100 kbps).

These have at best been temporary solutions on road to the high-speed broadband wireless experience that will be available on 3G.

Wireless videophones, high-speed internet access and TV will become a reality with 3G. Users may not watch a whole feature film on a mobile phone, but can view small clips.

A football match, cricket action or news will engage users. CD-quality music will ensure that the iPod will disappear behind a mobile phone.

People could end up reading most of their e-mails on mobiles and not on PCs. The big thing will be the 'banking of the unbanked' on the mobile phones. Micro payments will be possible via cell phones. The device will engulf the calendar, radio, MMS, video, TV, banking, camera, music & so on.

Some of these are already there in existing networks but slow transfer speeds limit the experience. Users will also be able to locate, navigate, and enhance security with GPS-assisted position services. That depends on digital maps available in public domain. A live video-conference will ensure that 3G users don't need to be in office.

Working of 3G

As every generation has been introduced with a new advanced technology, third generation (3G) is no exception. Latest developments in technology are not stunning now rather people are waiting impatiently for the newest modification and improvements and ready to go for it. 3G technology is the modified form of second generation it is better to say that all the best features of different versions of 2nd generation are combined into third generation. Resultantly 3G Technology is known as Smartphone with high data transfer rate, WiFi hotspots connectivity and multimedia features.

Three important technologies which make the 3G standard are CDMA 2000, WCDMA (UMTS), and TD-SCDMA. The prominent features after the integration of high capability technologies are like high speed data transfer rate at 3Mbps prop up the usage of internet. It's the matter of your need now to which mode you want to switch as to PC, internet, or phone mode, simply it's a 3 in 1. Let see the technologies used in 3G and their function how they work.

Working of 3G (CDMA 2000)

Code Division Multiple Access 2000 is approved by 3GPP2 Organization. CDMA200 hybrid with IS-95 B provides an unlimited access to IMT-200 Band as well as CDMA 200 1x and ideal conditions for the highest data transfer rate. The CDMA 2000 1x evolves into CDMA 200 1x EV. This cdma 200 1x EV IS put into service in two different forms:

- CDMA 2000 1x EV-DO- 1X Evolution data only able to use 1.25 MHz.

- CDMA 2000 1x EV-DV- 1x Evolution Data and Voice also use 1.25 MHz.

All these versions are supposed to attain the highest speed for greater efficiency of the mobile phones.

Working of 3G (W-CDMA / UMTS)

3G mobile technology has been marked by the CDMA accomplishment. ETSI Alpha group develop this technology on radio access method. W-CDMA offers challenges in shapes of versatility and complexity of its design. Its multifaceted single algorithm made the complete system more difficult hence the receiver becomes a more complex device. It provides friendly environment to the multi-users with greater simulation and broader interface able to transfer data with time variations.

UMTS network group is experimenting on this new technology to maintain previous 2G module features and the added features as well in 3G.

Working of how 3G (TD-SCDMA) Works

Developed by China Wireless Telecommunication Standard group TD-SCDMA is approved by the ITU. This technology is based on time synchronization with CDMA. This time division is based on duplex approach where uplink and downlink traffic transmit in different time slots. This synchronization gives flexibility to the spectrum for uplink and downlink transmission depending on symmetrical or asymmetrical information. Asymmetrical information comprises e-mail and internet applications while call system comes under symmetrical information. During asymmetrical applications downlink is given preference over uplink. Preference is given to uplink during telephony.

3G Mobile technology is providing 144 kbps connectivity speed which is the highest one in the present era. It is offering wireless broadband facility an entertainment opportunity of downloading music and videos, games with 3D effects and conference calls with video facility.

Based on the services, feature plans and areas with the 3G coverage there are many carriers who are providing it. It is still not fully accessible in all the countries so limitations hinder the availability. The real 3G technology can be enjoyed with a new mobile phone set and the 3G service pack.

All the three technologies mentioned above are working in 3G at their best.

4G or Fourth Generation

Fourth generation wireless (4G) is an abbreviation for the fourth generation of cellular wireless standards and replaces the third generation of broadband mobile communications. The standards for 4G, set by the radio sector of the International Telecommunication Union (ITU-R), are denoted as International Mobile Telecommunications Advanced (IMT-Advanced).

An IMT-Advanced cellular system is expected to securely provide mobile service users with bandwidth higher than 100 Mbps, enough to support high quality streaming multimedia content. Existing 3G technologies, often branded as Pre-4G (such as mobile WiMAX and 3G LTE), fall short of this bandwidth requirement. The majority of implementations branded as 4G do not comply with the full IMT-Advanced standard.

The premise behind the 4G service offering is to deliver a comprehensive IP based solution where multimedia applications and services can be delivered to the user anytime and anywhere with a high data rate, premium quality of service and high security.

Seamless mobility and interoperability with existing wireless standards is crucial to the functionality of 4G communications. Implementations will involve new technologies such as femtocell and picocell, which will address the needs of mobile users wherever they are and will free up network resources for roaming users or those in more remote service areas.

Two competing standards were submitted in September 2009 as technology candidates for ITU-R consideration:

- LTE Advanced - as standardized by the 3GPP.

- 802.16m - as standardized by IEEE.

These standards aim to be:

- Spectrally efficient.

- Able to dynamically allocate network resources in a cell.

- Able to support smooth handover.

- Able to offer high quality of service (QoS).

- Based on an all-IP packet-switched network.

WiMax is touted as the first 4G offering. It is an IP based, wireless broadband access technology, also known as IEEE 802.16. WiMax services offer residential and business customers with basic Internet connectivity.

Present implementations of WiMAX and LTE are largely considered a stopgap solution offering a considerable boost, while WiMAX 2 (based on the 802.16m specification) and LTE Advanced are finalized. Both technologies aim to reach the objectives traced by the ITU, but are still far from being implemented.

Enhanced Features of 4G Wireless Technology are as follows:

- Wider and extensive mobile coverage region.

- Larger bandwidth - higher data rates.

- Terminal Heterogeneity and Network Heterogeneity.

- Smoother and quicker handoff.

- WLAN for hot spots, an extension of 2G and 3G.

- Better scheduling and call admission control techniques.

- Global roaming and inter-working among various other access technologies.

- Supports interactive multimedia, video, wireless internet, voice and various other broadband services.

- User Friendliness and Personalization.

Benefits and Challenges

A. Benefits of 4G networks:

The benefits of 4G networks assist in ensuring a larger range of services and use-cases. However, the commercial models and eco-systems have not yet been established that are required in driving adoption from a user and service provider perspectives.

- Technology Performance Improvement: Delivers higher uplink and downlink throughput besides lesser latency and network capabilities. It has been universally believed that there will be a prolong growth in mobile data traffic significantly in the coming years. It is also a matter of fact that the majority of the core transport and throughput bottlenecks will undoubtedly be delivered by the technology itself despite of the 4G technology used (LTE or WiMAX) in comparison to 3G. 4G technologies provide at least two times more effective and efficient use of spectrum, enhanced support for real-time applications, and greater max speeds. Though there exist further network and capacity confronts such as edge or gateway management, signalling management which are needed to be fully addressed to increase benefits from the upgrade.

- New Mobile Application Enablement: It enables new mobile applications to enhance the existing ones (Streaming Music). Several 4G services such as digital storage or smart home monitoring will get enhanced by the improved 4G bandwidth and latency. Other services such as MMS, digital picture frames and various near-field communication applications will notice no significant improvement in riding on a 4G network. Hence, it is very crucial to have a very close look at the services and applications which are likely to be enhanced by 4G advancements. We can see that services which gain the most from the 4G technology's deployment are video streaming, MMOG/gaming and expertise applications such as interactive learning

- Addressable Device Expansion: Network potentials and chipset scale could expand the connectivity to various innovative gadgets. Handset technologies persist in evolving along a huge range of features and value added services by means of smart phones and more specialized gadgets. A carrier controlled service experience has been conventionally supported by the Terminal operating model. Commercial operating systems such as Windows Mobile or RIM have attracted heavy data users and hence fostered network congestion by reducing some control. In addition, the increasingly growing open eco-systems, further enabled by 4G, offer a challenging opportunity for operators since third parties develop services, applications and customization tools in order to meet user needs. Gadgets are becoming highly configurable because of open standards and more expertise gadgets such as net books, eReaders, tablets etc. are coming into the market. To meet lesser user segment needs we believe that vendors must think of a micro-segmentation based device roadmap; various new distribution channels are requisite to support the acceptation of Converged Mobile Gadgets and 4G applications.

- Differentiated Customer Experience: It enables in managing the user expectation and experience with new features and services. We consider the user's experience in gaining a profound understanding of how these services are completely facilitated and how it mingles into the fabric of our living, the necessity or capability to deploy expertise or configured gadgets to support enhancement, and finally, how to make money and when to share the income from the service delivery. Till now, it has been inadequate in understanding the experience of a 4G user and it is uncertain that how greatly the user experience will alter as many more and various 4G services arrive. We are much aware of the fact that user expectations regarding price points are retuning with growing expectations to pay "a little for a little" which confronts the present costing and monetization approaches. We also believe that users are expecting an additional bunching of services and applications into a "solution" which assists the way they live. Hence, accomplished adoption of 4G services will be highly reliant in resolving the most probable Use-Cases for 4G services.

- Business Model Evolution: 4G wireless technology will be the key in facilitatng the alternative partnership and monetization models. The previous couple of years or so have exposed the industry to the myth of all you can eat pricing models, or flat rate voice and data plans. This has motivated performance consistent with Pareto's data usage rule where 4% of users generally utilize more than 70% of the bandwidth. The consequential network bottlenecks restrain access in regions with a high tally of smart gadgets. The bandwidth requirements of several 4G use cases suggests that the above problem will only get worse if present pricing methods move further. One alternative presently being considered by operators motivates in moving towards the tiered pricing based on conventional aspects such as time, speed and quality of service. An additional capable service model is bandwidth on demand and the associated pricing method to charge premium pricing for these burst requirements. This may be proved advantageous in planning high bandwidth utilizing events such as video streaming or LIVE TV.

Given that what we are aware of today, 4G wireless technology will need an extension of pricing models to favour lower up-front prices (subscriptions, one time purchases, ad-based, fermium and per-use). Though, open development manifestoes and collaborative solution deployment/development methods may influence how manifold charging models may work. Undoubtedly new 4G service eco-system and use-cases arrangements head to the significant query of who will generate the bill for the services and how will the income be shared.

Challenges

Security and Privacy: Security measures must be instituted in the development of 4G Wireless Networks which will facilitate the safest possible technique for data transmission. Explicitly, "The 4G core delivers mobility, security, and QoS by means of reusing the existing methods while still working on a few mobility and handover concerns". Hence, for securing data, to be transmitted across the network, from hackers and further security contraventions it is obligatory for the organization to develop an efficient and effective series of tools which will support the utmost 4G security measures. As a result of the nature of the 4G wireless network, there is a more possibility of security intrusions, and hence, manifold levels of security, including increased necessities for validation, will be essential for protecting data and information transmitted across the network. One of the major objectives of 4G wireless networks is to envelope very wide geographic region with flawless service. Clearly, smaller local area networks will operate on different operating systems. The heterogeneity of these networks that exchange different sorts of data complicates the privacy and security concerns. Moreover, since new gadgets and services are being introduced for the first time in 4G wireless networks, the encryption and decryption schemes being used for 3G wireless networks are not suitable for 4G wireless networks. To prevail over these issues, two methods can be followed. The former method relies on modifying the current privacy and security methods so as to employ them to heterogeneous 4G wireless networks. The latter method relies in developing new, fresh dynamic reconfigurable, lightweight and adaptive mechanisms whenever the existing employed methods fail to get adapted to 4G wireless networks.

Quality of Service: Regarding the network quality, various telecommunication service providers assure the users for the enhanced connectivity, and the utmost possible data quality which is transmitted across the network, just as Ericsson's 4G Wireless Networks for TeliaSonera. With the data rates of almost 10 times higher as compared to today's conventional mobile broadband networks and real-time performance, it allows users to be connected always, even "on the move". Consequently, it is essential for service providers to develop an efficient and effective method to the 4G Wireless Networks which will improve quality, bestows effectual security measures, and will make sure that all users are provided with widespread options for downloading music, video, and picture files without any delays. The major confront for 4G wireless networks is incorporating IP-based and non-IP-based gadgets. We know that gadgets which are non-IP address based are usually used for services such as VoIP. In contrast, gadgets which are IP address based are generally used for delivering data.

Working of 4G

4G works much in the same way as 3G, simply faster. Using high-speed download and upload packets, 4G allows you to access broadband style speeds whilst away from your Wi-Fi. Users can often access speeds of up to 21Mb on the go, but this is, however, affected by location. A larger city, for example, will exhibit faster speeds than a small village.

4G is essentially a highly advanced radio system. You may even have seen masts dotted around the landscape. These masts broadcast the signals necessary for 4G to work and the challenge is for engineers and coders is to cram as much data into these signals as possible. By extension, this means the network is faster and more efficient.

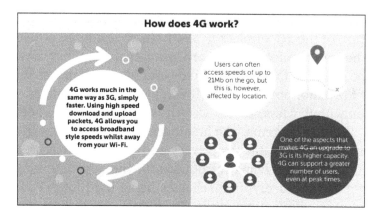

Like 3G, 4G is a protocol that sends and receives data in packets. However, 4G differs from 3G in how it works. 4G is entirely IP based, which means it uses internet protocols even for voice data. Conforming to this one standard means it is less likely for data to become scrambled while traversing the various networks, meaning a more seamless experience for us users.

Like all mobile broadband, 4G works through your device communicating with a base station. Base stations are technical speak for the masts that we've all seen popping up throughout the country. This mast relays data from your device to the internet and back again.

One of the aspects that makes 4G an upgrade to 3G is its higher capacity. 4G can support a greater number of users, even at peak times. For example, a 3G tower may only be able to give 100 people the best possible connection speed, but a 4G tower can theoretically give 400 people the best service.

4G also features reduced latency, which if you're a mobile gamer is essential. With reduced latency, you'll see a much quicker response to your commands. So for gamers, if they are playing a fighting game, for example, this can mean the difference between winning and losing.

Once 4G infrastructures become more common, users will see more seamless streaming on the move from services like YouTube, better video calls and even better battery life. At the moment 4G signals are rarer outside of big cities, so phones expend a lot of energy looking around for a 4G signal.

5G or Fifth Generation

5G Technology stands for 5th Generation Mobile technology. 5G mobile technology has changed the means to use cell phones within very high bandwidth. User never experienced ever before such a high value technology. Nowadays mobile users have much awareness of the cell phone (mobile) technology. The 5G technologies include all type of advanced features which makes 5G mobile technology most powerful and in huge demand in near future.

The gigantic array of innovative technology being built into new cell phones is stunning. 5G technology which is on hand held phone offering more power and features than at least 1000 lunar modules. A user can also hook their 5G technology cell phone with their Laptop to get broadband internet access. 5G technology including camera, MP3 recording, video player, large phone memory, dialing speed, audio player and much more you never imagine. For children rocking fun Bluetooth technology and Pico nets has become in market.

5G technology going to be a new mobile revolution in mobile market. Through 5G technology now you can use worldwide cellular phones and this technology also strike the china mobile market and a user being proficient to get access to Germany phone as a local phone. With the coming out of cell phone alike to PDA now your whole office in your finger tips or in your phone. 5G technology has extraordinary data capabilities and has ability to tie together unrestricted call volumes and infinite data broadcast within latest mobile operating system. 5G technology has a bright future because it can handle best technologies and offer priceless handset to their customers. May be in coming days 5G technology takes over the world market. 5G Technologies have an extraordinary capability to support Software and Consultancy. The Router and switch technology used in 5G network providing high connectivity. The 5G technology distributes internet access to nodes within the building and can be deployed with union of wired or wireless network connections. The current trend of 5G technology has a glowing future.

5G network is assumed as the perfection level of wireless communication in mobile technology. Cable network is now become the memory of past. Mobiles are not only a communication tool but also serve many other purposes. All the previous wireless technologies are entertaining the ease of telephone and data sharing but 5G is bringing a new touch and making the life real mobile life. The new 5G network is expected to improve the services and applications offered by it.

5G network is very fast and reliable. The concept of handheld devices is going to be revolutionized with the advent of 5G. Now all the services and applications are going to be accessed by single IP as telephony, gaming and many other multimedia applications. As it is not a new thing or gadget in market and there are millions of thousands users all over the world who have experienced the wireless services and till now they are addicted to this wireless technology. It is not easy for them to shrink from using this new 5G network technology. There is only need to make it accessible so that a common man can easily afford the economic packs offered by the companies so that 5G network could hold the authentic place. There is need to win the customer trust to build fair long term relation to make a reliable position in the telecommunication field.

To compete with the preceding wireless technologies in the market 5G network has to offer something reliable something more innovative. All the features like telephony, camera, mp3 player, are coming in new mobile phone models. 4G is providing all these utility in mobile phone. By seeing the features of 4G one can gets a rough idea about what could be a 5g Network offer. There is messenger, photo gallery, and multimedia applications that are also going to be the part of 5G. There would be no difference between a PC and a mobile phone rather both would act vice versa.

5G Network Features

A first remarkable feature of 5G network is the broadband internet in mobile phones that would be possible to provide internet facility in the computer by just connecting the mobile.

5G technology offer high resolution for crazy cell phone user and bi-directional large bandwidth shaping:

- The advanced billing interfaces of 5G technology makes it more attractive and effective.

- 5G technology also providing subscriber supervision tools for fast action.

- The high quality services of 5G technology based on Policy to avoid error.

- 5G technology is providing large broadcasting of data in Gigabit which supporting almost 65,000 connections.

- 5G technology offer transporter class gateway with unparalleled consistency.

- The traffic statistics by 5G technology makes it more accurate.

- Through remote management offered by 5G technology a user can get better and fast solution.

- The remote diagnostics also a great feature of 5G technology.

- The 5G technology is providing up to 25 Mbps connectivity speed.

- The 5G technology also support virtual private network.

- The new 5G technology will take all delivery service out of business prospect.

- The uploading and downloading speed of 5G technology touching the peak.

- The 5G technology network offering enhanced and available connectivity just about the world.

A new revolution of 5G technology is about to begin because 5G technology going to give tough completion to normal computer and laptops whose marketplace value will be effected. There are lots of improvements from 1G, 2G, 3G, and 4G to 5G in the world of telecommunications. The new coming 5G technology is available in the market in affordable rates, high peak future and much reliability than its preceding technologies.

Working of 5G

Like other cellular networks, 5G networks use a system of cell sites that divide their territory into sectors and send encoded data through radio waves. Each cell site must be connected to a network backbone, whether through a wired or wireless backhaul connection.

5G networks use a type of encoding called OFDM, which is similar to the encoding that 4G LTE uses. The air interface is designed for much lower latency and greater flexibility than LTE, though.

With the same airwaves as 4G, the 5G radio system can get about 30 percent better speeds thanks to more efficient encoding. The crazy gigabit speeds you hear about are because 5G is designed to use much larger channels than 4G does. While most 4G channels are 20MHz, bonded together into up to 160MHz at a time, 5G channels can be up to 100MHz, with Verizon using as much as 800MHz at a time. That's a much broader highway, but it also requires larger, clear blocks of airwaves than were available for 4G.

That's where the higher, short-distance millimeter-wave frequencies come in. While lower frequencies are occupied—by 4G, by TV stations, by satellite firms, or by the military—there had been a huge amount of essentially unused higher frequencies available in the US, so carriers could easily construct wide roads for high speeds.

5G networks need to be much smarter than previous systems, as they're juggling many more, smaller cells that can change size and shape. But even with existing macro cells, Qualcomm says 5G will be able to boost capacity by four times over current systems by leveraging wider bandwidths and advanced antenna technologies.

The goal is to have far higher speeds available, and far higher capacity per sector, at far lower latency than 4G. The standards bodies involved are aiming at 20Gbps speeds and 1ms latency, at which point very interesting things begin to happen.

LTE Network Architecture

LTE stands for Long Term Evolution and it was started as a project in 2004 by telecommunication body known as the Third Generation Partnership Project (3GPP). SAE (System Architecture Evolution) is the corresponding evolution of the GPRS/3G packet core network evolution. The term LTE is typically used to represent both LTE and SAE.

LTE evolved from an earlier 3GPP system known as the Universal Mobile Telecommunication System (UMTS), which in turn evolved from the Global System for Mobile Communications (GSM). Even related specifications were formally known as the evolved UMTS terrestrial radio access (E-UTRA) and evolved UMTS terrestrial radio access network (E-UTRAN). First version of LTE was documented in Release 8 of the 3GPP specifications.

A rapid increase of mobile data usage and emergence of new applications such as MMOG (Multimedia Online Gaming), mobile TV, Web 2.0, streaming contents have motivated the 3rd Generation Partnership Project (3GPP) to work on the Long-Term Evolution (LTE) on the way towards fourth-generation mobile.

The main goal of LTE is to provide a high data rate, low latency and packet optimized radioaccess technology supporting flexible bandwidth deployments. Same time its network architecture has been designed with the goal to support packet-switched traffic with seamless mobility and great quality of service.

LTE Evolution

Year	Event
Mar 2000	Release 99 - UMTS/WCDMA.
Mar 2002	Rel 5 - HSDPA.
Mar 2005	Rel 6 - HSUPA.
Year 2007	Rel 7 - DL MIMO, IMS (IP Multimedia Subsystem).
November 2004	Work started on LTE specification.
January 2008	Spec finalized and approved with Release 8.
2010	Targeted first deployment.

- LTE is the successor technology not only of UMTS but also of CDMA 2000.

- LTE is important because it will bring up to 50 times performance improvement and much better spectral efficiency to cellular networks.

- LTE introduced to get higher data rates, 300Mbps peak downlink and 75 Mbps peak uplink. In a 20MHz carrier, data rates beyond 300Mbps can be achieved under very good signal conditions.

- LTE is an ideal technology to support high date rates for the services such as voice over IP (VOIP), streaming multimedia, videoconferencing or even a high-speed cellular modem.

- LTE uses both Time Division Duplex (TDD) and Frequency Division Duplex (FDD) mode. In FDD uplink and downlink transmission used different frequency, while in TDD both uplink and downlink use the same carrier and are separated in Time.

- LTE supports flexible carrier bandwidths, from 1.4 MHz up to 20 MHz as well as both FDD and TDD. LTE designed with a scalable carrier bandwidth from 1.4 MHz up to 20 MHz which bandwidth is used depends on the frequency band and the amount of spectrum available with a network operator.

- All LTE devices have to support (MIMO) Multiple Input Multiple Output transmissions, which allow the base station to transmit several data streams over the same carrier simultaneously.

- All interfaces between network nodes in LTE are now IP based, including the backhaul connection to the radio base stations. This is great simplification compared to earlier technologies that were initially based on E1/T1, ATM and frame relay links, with most of them being narrowband and expensive.

- Quality of Service (QoS) mechanism have been standardized on all interfaces to ensure that the requirement of voice calls for a constant delay and bandwidth, can still be met when capacity limits are reached.

- Works with GSM/EDGE/UMTS systems utilizing existing 2G and 3G spectrum and new spectrum. Supports hand-over and roaming to existing mobile networks.

Advantages of LTE

- High throughput: High data rates can be achieved in both downlink as well as uplink. This causes high throughput.

- Low latency: Time required to connect to the network is in range of a few hundred milliseconds and power saving states can now be entered and exited very quickly.

- FDD and TDD in the same platform: Frequency Division Duplex (FDD) and Time Division Duplex (TDD), both schemes can be used on same platform.

- Superior end-user experience: Optimized signaling for connection establishment and other air interface and mobility management procedures have further improved the user experience. Reduced latency (to 10 ms) for better user experience.

- Seamless Connection: LTE will also support seamless connection to existing networks such as GSM, CDMA and WCDMA.

- Plug and play: The user does not have to manually install drivers for the device. Instead system automatically recognizes the device, loads new drivers for the hardware if needed, and begins to work with the newly connected device.

- Simple architecture: Because of Simple architecture low operating expenditure (OPEX).

LTE - QoS

LTE architecture supports hard QoS, with end-to-end quality of service and guaranteed bit rate (GBR) for radio bearers. Just as Ethernet and the internet have different types of QoS, for example, various levels of QoS can be applied to LTE traffic for different applications. Because the LTE MAC is fully scheduled, QoS is a natural fit.

Evolved Packet System (EPS) bearers provide one-to-one correspondence with RLC radio bearers and provide support for Traffic Flow Templates (TFT). There are four types of EPS bearers:

- GBR Bearer resources permanently allocated by admission control.

- Non-GBR Bearer no admission control.

- Dedicated Bearer associated with specific TFT (GBR or non-GBR).

- Default Bearer Non GBR, catch-all for unassigned traffic.

LTE Network Architecture

The high-level network architecture of LTE is comprised of following three main components:

- The User Equipment (UE).

- The Evolved UMTS Terrestrial Radio Access Network (E-UTRAN).

- The Evolved Packet Core (EPC).

The evolved packet core communicates with packet data networks in the outside world such as the internet, private corporate networks or the IP multimedia subsystem. The interfaces between the different parts of the system are denoted Uu, S1 and SGi as shown below:

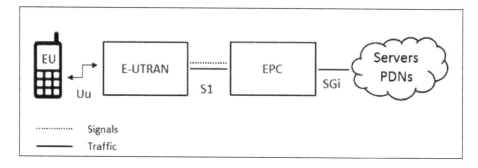

The User Equipment (UE)

The internal architecture of the user equipment for LTE is identical to the one used by

UMTS and GSM which is actually a Mobile Equipment (ME). The mobile equipment comprised of the following important modules:

- Mobile Termination (MT): This handles all the communication functions.

- Terminal Equipment (TE): This terminates the data streams.

- Universal Integrated Circuit Card (UICC): This is also known as the SIM card for LTE equipments. It runs an application known as the Universal Subscriber Identity Module (USIM).

A USIM stores user-specific data very similar to 3G SIM card. This keeps information about the user's phone number, home network identity and security keys etc.

The E-UTRAN (The Access Network)

The architecture of evolved UMTS Terrestrial Radio Access Network (E-UTRAN) has been illustrated below.

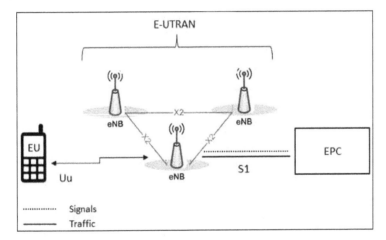

The E-UTRAN handles the radio communications between the mobile and the evolved packet core and just has one component, the evolved base stations, called eNodeB or eNB. Each eNB is a base station that controls the mobiles in one or more cells. The base station that is communicating with a mobile is known as its serving eNB.

LTE Mobile communicates with just one base station and one cell at a time and there are following two main functions supported by eNB:

- The eBN sends and receives radio transmissions to all the mobiles using the analogue and digital signal processing functions of the LTE air interface.

- The eNB controls the low-level operation of all its mobiles, by sending them signalling messages such as handover commands.

Each eBN connects with the EPC by means of the S1 interface and it can also be connected to nearby base stations by the X2 interface, which is mainly used for signalling and packet forwarding during handover.

A home eNB (HeNB) is a base station that has been purchased by a user to provide femtocell coverage within the home. A home eNB belongs to a closed subscriber group (CSG) and can only be accessed by mobiles with a USIM that also belongs to the closed subscriber group.

The Evolved Packet Core (EPC): The Core Network

The architecture of Evolved Packet Core (EPC) has been illustrated below. There are few more components which have not been shown in the diagram to keep it simple. These components are like the Earthquake and Tsunami Warning System (ETWS), the Equipment Identity Register (EIR) and Policy Control and Charging Rules Function (PCRF).

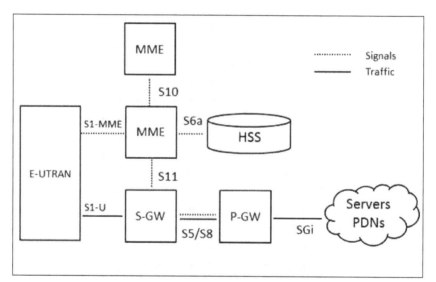

Below is a brief description of each of the components shown in the above architecture:

- The Home Subscriber Server (HSS) component has been carried forward from UMTS and GSM and is a central database that contains information about all the network operator's subscribers.

- The Packet Data Network (PDN) Gateway (P-GW) communicates with the outside world ie. packet data networks PDN, using SGi interface. Each packet data network is identified by an access point name (APN). The PDN gateway has the same role as the GPRS support node (GGSN) and the serving GPRS support node (SGSN) with UMTS and GSM.

- The serving gateway (S-GW) acts as a router, and forwards data between the base station and the PDN gateway.

- The mobility management entity (MME) controls the high-level operation of the mobile by means of signalling messages and Home Subscriber Server (HSS).

- The Policy Control and Charging Rules Function (PCRF) is a component which is not shown in the above diagram but it is responsible for policy control decision-making, as well as for controlling the flow-based charging functionalities in the Policy Control Enforcement Function (PCEF), which resides in the P-GW.

The interface between the serving and PDN gateways is known as S5/S8. This has two slightly different implementations, namely S5 if the two devices are in the same network, and S8 if they are in different networks.

Functional Split between the E-UTRAN and the EPC

Following diagram shows the functional split between the E-UTRAN and the EPC for an LTE network:

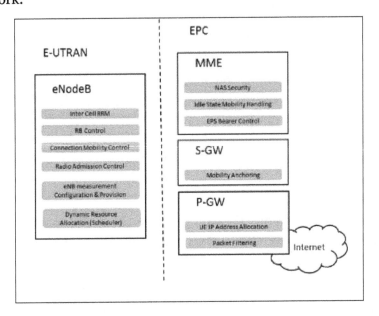

2G/3G versus LTE

Following table compares various important Network Elements & Signaling protocols used in 2G/3G abd LTE.

2G/3G	LTE
GERAN and UTRAN	E-UTRAN
SGSN/PDSN-FA	S-GW
GGSN/PDSN-HA	PDN-GW
HLR/AAA	HSS
VLR	MME

SS7-MAP/ANSI-41/RADIUS	Diameter
DiameterGTPc-v0 and v1	GTPc-v2
MIP	PMIP

A network run by one operator in one country is known as a Public Land Mobile Network (PLMN) and when a subscribed user uses his operator's PLMN then it is said Home-PLMN but roaming allows users to move outside their home network and using the resources from other operator's network. This other network is called Visited-PLMN.

A roaming user is connected to the E-UTRAN, MME and S-GW of the visited LTE network. However, LTE/SAE allows the P-GW of either the visited or the home network to be used, as shown in below:

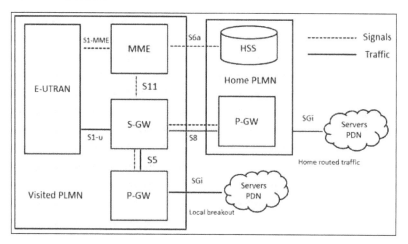

The home network's P-GW allows the user to access the home operator's services even while in a visited network. A P-GW in the visited network allows a "local breakout" to the Internet in the visited network.

The interface between the serving and PDN gateways is known as S5/S8. This has two slightly different implementations, namely S5 if the two devices are in the same network, and S8 if they are in different networks. For mobiles that are not roaming, the serving and PDN gateways can be integrated into a single device, so that the S5/S8 interface vanishes altogether.

LTE Roaming Charging

The complexities of the new charging mechanisms required to support 4G roaming are much more abundant than in a 3G environment. Few words about both pre-paid and post-paid charging for LTE roaming is given below:

- Prepaid Charging: The CAMEL standard, which enables prepaid services in 3G, is not supported in LTE; therefore, prepaid customer information must be routed back to the home network as opposed to being handled by the local visited network. As a result, operators must rely on new accounting flows to access prepaid customer data, such as through their P-Gateways in both IMS and non-IMS environments or via their CSCF in an IMS environment.

- Postpaid Charging: Postpaid data-usage charging works the same in LTE as it does in 3G, using versions TAP 3.11 or 3.12. With local breakout of IMS services, TAP 3.12 is required.

Operators do not have the same amount of visibility into subscriber activities as they do in home-routing scenarios in case of local breakout scenarios because subscriber-data sessions are kept within the visited network; therefore, in order for the home operator to capture real-time information on both pre- and postpaid customers, it must establish a Diameter interface between charging systems and the visited network's P-Gateway.

In case of local breakout of ims services scenario, the visited network creates call detail records (CDRs) from the S-Gateway(s), however, these CDRs do not contain all of the information required to create a TAP 3.12 mobile session or messaging event record for the service usage. As a result, operators must correlate the core data network CDRs with the IMS CDRs to create TAP records.

References

- What-is-mobile-radio-telephone-system-or-0g-and-what-is-0-5g, technology: cleardoubts.com, Retrieved 06 August, 2019

- What-is-1g-or-first-generation-of-wireless-telecommunication-technology, technology: cleardoubts.com, Retrieved 09 January, 2019

- 2g-technology: freewimaxinfo.com, Retrieved 25 July, 2019

- What-is-3g-technology-all-about, industry-telecom -1678989: economictimes.indiatimes.com, Retrieved 04 April, 2019

- Forth-generation-wireless-4g- 2920: techopedia.com, Retrieved 18 June, 2019

- 5g-technology: freewimaxinfo.com, Retrieved 14 April, 2019

- Lte-overview, lte: tutorialspoint.com, Retrieved 08 January, 2019

Mobile Network Security

Mobile communication network security is an essential component in mobile computing as it provides security of personal and business information stored in smartphones. Some of its aspects are GSM security, CDMA security, 3G security, 4G and LTE network security. This chapter discusses these aspects related to mobile communication network security in detail.

The increasing use of mobile communication networks results in ever more stringent security requirements. In an information society, availability, integrity and confidentiality are essential. Especially the provision of the latter is hard to demonstrate. If someone or some component is able to collect and store personal data, one cannot be sure that this data is not gathered and not misused. But this "being sure" is essential with respect to privacy and data protection.

Threats on Cellular Networks

Unlike wired networks, mobile cellular networks have a complex structure with many components and protocols resulting in vulnerabilities and threats. However, it is possible to take the necessary precautions and thus mitigate the risks of these threats to an acceptable level. Mobile operators should take the required actions to mitigate the security risks to their subscribers and also to their core networks. It is up to an operator to possibly charge for these security actions.

In mobile cellular networks, paging effects are important security issues. Paging attacks could be most damaging on the border of a mobile operator's network. One border of cellular networks is where Internet access starts and another one is where subscribers attach to a 3G network. Mobile operators can take most of the countermeasures on these border points. Thanks to the release of the 3G and LTE technologies, mobile GSM operators have overtaken ISP roles. Some argue that, like traditional ISPs, mobile operators have no obligation to protect their customers, while others argue they should protect their subscribers, as well as their core network infrastructure, against security threats. Although traditional wired network ISPs and mobile network operators both act as ISPs, their core network infrastructure or underlying networks are very different. The security requirement on the border of the Internet and on the subscriber border is higher for mobile operators than for traditional wired ISPs. This is due to the existence of more fragile components and protocols used in mobile networks. We've focused on these border points. In an actual production mobile network environment, we have investigated the effect of DoS kind of attacks from the Internet against 3G users.

User Data and Switching Data

Data to be protected when communicating can be subdivided in user data and switching data. User data is data given to the network to be transmitted. Switching data is data needed for connecting sender and recipient. Data of content, data of interest and traffic data can be derived from user and switching data.

In digital mobile systems the implementation of encryption is inexpensive because of digital data transmission and could therefore be easily applied. But, for example, systems of the GSM standard use only link-to-link encryption at the radio interface to protect user data. If encryption algorithms are not publicly known nobody can be sure that data cannot be read by third parties and/or data of interest can be filtered out. Obviously, the network operator has this possibility since he creates the encryption keys. For example, in GSM switching data is stored in registers (home location register HLR, visitor location register VLR). This data has to be transmitted for billing. This data exchange cannot easily be traced especially if network operators and service providers are not the same institution.

If such data is accumulated it allows conclusions about the interests of network users. It can tell who has communicated with whom how long. Traffic data allow to construct location profiles.

Requirements Resulting from Data Protection

The ability to collect and analyze data of others on a large scale, we consider as a violation of personal privacy – since there are possibilities to design and implement communication systems in such a way that they do not enable this.

For public mobile communication systems intended for broad use, in our opinion, the following requirements resulting from data protection should be met:

Protection of Confidentiality:

- Message contents should be kept confidential towards all parties except the communication partners.

- Sender and/or addressee of messages should stay anonymous to each other, and third parties (including the network operator(s)) should be unable to observe their communication.

- Neither potential communication partners nor third parties (including the network operator(s)) should be able to locate mobile stations or their users.

Protection of Integrity:

- Forging message contents (including sender's address) should be detected.

- The recipient of a message y should be able to prove to third parties that entity x has sent message y

- The sender of a message should be able to prove the sending of a message with correct contents, if possible, even that the addressee received the message.

- Nobody can cheat the network operator(s) with in terms of usage fees. But on the other hand, the network operator(s) can only demand usage fees for correctly delivered services.

Protection of Availability:

- The communication network enables communication between all parties who wish to communicate (and who are allowed to).

Such data protection requirements cannot be fulfilled by legal means alone – and laws, which cannot be enforced, have a negative effect of law-abiding in general. Confidentiality requirements must therefore be enforced by the prevention of the gathering of personal data. There is no other way known to achieve privacy with respect to the operator and designer of network components.

With respect to the latter, it is mostly completely ignored that a component or system might be under control of everybody who has had access to it so far, because he might have installed a Trojan Horse. Not only the designer but also every complex tool used to design the system might be able to do so. Moreover, the Trojan Horse in the first tool may be implemented by another tool used to generate the first one and so on (transitive Trojan Horse).

Realization of Data Protection Requirements

The fact of mobility makes it difficult to apply well known concepts in the same way as in fixed networks. After outlining these basic concepts some ideas are given which could point into the right direction.

Basic Concepts

Security problems may be solved by using methods such as end-to-end-encryption and linkto-link-encryption. The anonymity of participants can be protected by using certain kinds of addressing, broadcast and other methods for example MIXes and superposed sending.

Protection of User Data

Requirement c_1 means, trusted communication between two participants of the same and of other networks must be possible. The same must be true for integrity requirements i_1, i_2 and i_3. These can be achieved by encryption, digital signatures and authentication codes. In fact, c_1 can be accomplished by end-to-end-encryption. For i_1 to be realized, for example, a hash-value of a message is digitally signed. For i3 a digital signature of the sender of the message is necessary. Fulfillment of requirement i3 needs a signed receipt from the recipient or the message transmission system.

Cryptography is only applicable if the following conditions are true:

- The different services and different network systems need to match corresponding encryption methods and protocols. But this seems to be more a legal (political) and economical than a technical problem.

- User channels (according to the OSI-7-layer-model of the ISO at the transport layer 4 or higher) must be bit-transparent, i.e. bits to be transmitted on the signal path must not be changed or interfered with. The minor change of bits could mean a loss of integrity on the signal path. Furthermore, a change of only one bit would be followed by an increased rate of errors since encryption systems usually produce a strong dependency between bits.

- Even bit-transparency is not implemented in every already realized and standardized network, e.g. systems of the GSM standard have non bit-transparent speech channels but in network systems like ISDN bit-transparent channels are available. Considering these aspects the integration of networks and services must be planned carefully.

Protection of Switching Data

The following concepts show the possibility of developing networks which fulfill our data protection requirements.

Link-to-Link Encryption

The contents of a message can be sufficiently hidden by end-to-end encryption at the ISO layer 4. If protocols of the layers 1 to 3 also contain personal data then it is also necessary to protect this information by link-to-link encryption. This information could be the address of a mobile terminal or the address of a smart card. This data is strongly related to the owner because such equipment is of a very personal character and will not be changed after every usage.

Recipient Anonymity by Broadcast and Addressing Attributes

Receiving a message can be made completely anonymous to the network by delivering the message (possibly end-to-end-encrypted) to all stations (broadcast). If the message has an intended recipient, a so called addressee, it has to contain an attribute by which he and nobody else can recognize it as addressed to him. This attribute is called an implicit address. It is meaningless and only understandable by the recipient who can determine whether he is the intended person. In contrast, an explicit address describes either a place in the network or the place of a station.

Implicit addresses can be distinguished according to their visibility, i.e. whether they can be tested for equality or not. An implicit address is called invisible, if it is only visible to its addressee and is called visible otherwise.

implicit address		address distribution	
		public address	private address
	invisible	very costly, but necessary to establish contact	costly
	visible	not advisable	frequent change after use

Combination of implicit addressing modes and address distribution.

Invisible implicit addresses, unfortunately very costly, can be realized with a public key cryptosystem. Visible implicit addresses can be realized much easier. Users choose arbitrary names for themselves, which can then be prefixed to messages.

Another criterion to distinguish implicit addresses is their distribution. An implicit address is called public, if it is known to every user (like telephone numbers today) and private if the sender received it secretly from the addressee either outside the network or as a return address or by a generating algorithm the sender and the addressee agreed upon.

Public addresses should not be realized by visible implicit addresses to avoid the linkability of the visible public address of a message and the addressed user.

Private addresses can be realized by visible addresses but then each of them should be used only once.

Sender Anonymity by using DC-network or MIX-network

A powerful scheme for sender anonymity is superposed sending which is published in and is called DC-network (dining cryptographers network) there. For DC-networks it is proved that the anonymity of the sender is protected as long as station are linked by exchanged keys unknown to the attacker.

Unlinkability of sender and recipient can be realized by a special network station, a so called MIX, which collects a number of messages of equal length from many distinct senders, discards repeats, changes their encodings, and forwards the messages to the recipients in a different order. This measure hides the relation between sender and recipient of a message from everybody but the MIX and the sender of the message. Change of encoding of a message can be implemented using a public-key cryptosystem. Since decryption is a deterministic operation, repeats of messages have to be discarded. Otherwise, the change of encoding does not prevent tracing messages through MIXes. Simply count the number of copies of each message before and after the MIX.

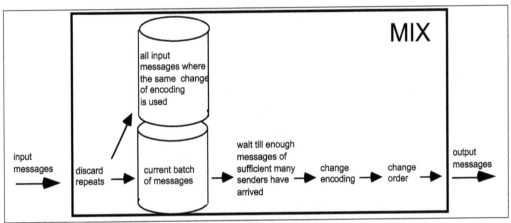

Functionality of MIXes.

By using more than one MIX to forward a message from the sender to the recipient, the relation is hidden from all attackers in the network who do not control all MIXes which the message passed, nor have the cooperation of the sender. MIXes should be independently designed and produced and should have independent operators, otherwise a single party is able to control a communication.

One method of achieving sender anonymity, i.e. of making unclear when a message was sent, is dummy-traffic. That means each user station sends among the real messages a number of meaningless messages.

Unfortunately the above described concepts are hardly feasible for the protection of the radio interface, e.g. dummy-traffic is not acceptable due to limited accumulator capacity and bandwidth. But, most of the concepts to protect switching data and interest data can be realized at the part of a communication system which is the fixed network.

Possibilities for Anonymous and Secure Accounting

In a public network there should exist a convenient and secure way to handle fees for used services . Accounting mechanisms should be organized in a manner that anonymity and unobservability in communication networks are guaranteed.

In principle, there are two possible approaches: individual accounting and flat-rate accounting. The latter can be handled freely, i.e. without taking care of anonymity because no interesting subscriber-specific information is needed.

By using flat-rate accounting problems of fraud will be avoided. Furthermore, there is no need for a complex accounting management system. If there is enough bandwidth, as this is the case in fixed networks, flat-rate accounting can be used for local calls.

Individual accounting means to apply methods which preserve the anonymity of the person. It may be accomplished by installing an unmanipulatable accounting system placed at a fixed location combined with a mobile part. For example the fixed part can be placed at home and the mobile part in a mobile unit known as SIM (Subscribers Identity Module). Accounting can also be done by digital paying on-line towards the fixed part or by first buying a smart card like a pre-paid telephone card, that accumulated a number of service to be used. Main advantage of an unmanipulatable accounting system is, that the network does not need to manage accounting. Otherwise, manipulation at the fixed accounting equipment and at the mobile equipment could be done by the subscriber. These manipulation or a change of the accumulated number on the SIM must be recognized by network operator. Another approach of individual accounting is the use of an anonymous digital accounting system . This guarantees anonymity and unobservability. It is important to design the protocol outlines in a way that nobody can use the system in a fraudulent way because anonymous digital accounting systems prevent prosecution after fraud took place.

Security Issues in Cellular Networks

The infrastructure for Cellular Networks is massive, complex with multiple entities co-ordinating together, such as the IP Internet coordinating with the core network. And therefore it presents a challenge for the network to provide security at every possible communication path.

Limitations of Cellular Networks

Compared to Wired Networks, Wireless Cellular Networks have a lot of limitations.

- Open Wireless Access Medium: Since the communication is on the wireless channel, there is no physical barrier that can separate an attacker from the network.

- Limited Bandwidth: Although wireless bandwidth is increasing continuously, because of channel contention everyone has to share the medium.

- System Complexity: Wireless systems are more complex due to the need to support mobility and making use of the channel effectively. By adding more complexity to systems, potentially new security vulnerabilities can be introduced.

- Limited Power: Wireless Systems consume a lot of power and therefore have a limited time battery life.

- Limited Processing Power: The processors installed on the wireless devices are increasing in power, but still they are not powerful enough to carry out intensive processing.

- Relatively Unreliable Network Connection: The wireless medium is an unreliable medium with a high rate of errors compared to a wired network.

There are several security issues that have to be taken into consideration when deploying a cellular infrastructure. The importance of which has increased with the advent of advanced networks like 3G.

- Authentication: Cellular networks have a large number of subscribers, and each has to be authenticated to ensure the right people are using the network. Since the purpose of 3G is to enable people to communicate from anywhere in the world, the issue of cross region and cross provider authentication becomes an issue.

- Integrity: With services such as SMS, chat and file transfer it is important that the data arrives without any modifications.

- Confidentiality: With the increased use of cellular phones in sensitive communication, there is a need for a secure channel in order to transmit information.

- Access Control: The Cellular device may have files that need to have restricted access to them. The device might access a database where some sort of role based access control is necessary.

- Operating Systems In Mobile Devices: Cellular Phones have evolved from low processing power, ad-hoc supervisors to high power processors and full fledged operating systems. Some phones may use a Java Based system, others use Microsoft Windows CE and have the same capabilities as a desktop computer. Issues may arise in the OS which might open security holes that can be exploited.

- Web Services: A Web Service is a component that provides functionality accessible through the web using the standard HTTP Protocol. This opens the cellular device to variety of security issues such as viruses, buffer overflows, denial of service attacks etc.

- Location Detection: The actual location of a cellular device needs to be kept hidden for reasons of privacy of the user. With the move to IP based networks, the issue arises that a user may be associated with an access point and therefore their location might be compromised.

- Viruses And Malware: With increased functionality provided in cellular systems, problems prevalent in larger systems such as viruses and malware arise. The first virus that appeared on cellular devices was Liberty. An affected device can also be used to attack the cellular network infrastructure by becoming part of a large scale denial of service attack.

- Downloaded Contents: Spyware or Adware might be downloaded causing security issues. Another problem is that of digital rights management. Users might download unauthorized copies of music, videos, wallpapers and games.

- Device Security: If a device is lost or stolen, it needs to be protected from unauthorized use so that potential sensitive information such as emails, documents, phone numbers etc. cannot be accessed.

Types of Attacks

Due to the massive architecture of a cellular network, there are a variety of attacks that the infrastructure is open to.

- Denial Of Service (DOS): This is probably the most potent attack that can bring down the entire network infrastructure. This is caused by sending excessive data to the network, more than the network can handle, resulting in users being unable to access network resources.

- Distributed Denial Of Service (DDOS): It might be difficult to launch a large scale DOS attack from a single host. A number of hosts can be used to launch an attack.

- Channel Jamming: Channel jamming is a technique used by attackers to jam the wireless channel and therefore deny access to any legitimate users in the network.

- Unauthorized Access: If a proper method of authentication is not deployed then an attacker can gain free access to a network and then can use it for services that he might not be authorized for.

- Eavesdropping: If the traffic on the wireless link is not encrypted then an attacker can eavesdrop and intercept sensitive communication such as confidential calls, sensitive documents etc.

- Message Forgery: If the communication channel is not secure, then an attacker can intercept messages in both directions and change the content without the users ever knowing.

- Message Replay: Even if communication channel is secure, an attacker can intercept an encrypted message and then replay it back at a later time and the user might not know that the packet received is not the right one.

- Man In The Middle Attack: An attacker can sit in between a cell phone and an access station and intercept messages in between them and change them.

- Session Hijacking: A malicious user can highjack an already established session, and can act as a legitimate base station.

Measuring the Effect of the DoS Attack on 3G Networks

Paging is a big threat in mobile cellular networks. So, we performed some attacks by exploiting vulnerabilities in GSM network components due to paging, and examined the effect of these attacks. A network or company is as strong as its weakest point in terms of security. Therefore, a network should be assessed as a whole when a security solution is put in place. All pieces of a network should be evaluated carefully in terms of security. We performed DoS and flooding attacks in a production mobile cellular network that supports 120,000 concurrent users and investigated their effect on core network equipment such as the RNC and SGSN. We suggest some countermeasures, taking our findings into consideration.

Some mobile cellular operators think that no security solution is necessary on the Internet border of their network that serves their subscribers during Internet access. So, they typically construct their networks as shown in Figure. In this setting, any packet coming from the Internet, destined to a subscriber of the mobile operator, could reach its target.

A subscriber accesses core network devices for authentication and authorization, after passing through the air interface and the transport network of the mobile operator. An IP address is assigned to a subscriber by a GGSN if it is successfully authenticated and authorized. The first point on the mobile cellular network where a subscriber can do IP-based communication is the GGSN. The GGSN acts as a remote access server. A subscriber's data traffic is carried via tunnels until the GGSN. For example, the subscriber traffic from the GGSN to the SGSN is carried by GTP tunnels. Briefly, behind the GGSN, the subscriber data traffic is tunneled and carried to the subscriber as shown in figure. Therefore, when any customer tries to learn the traffic path by using trace applications (e.g., tracert for Windows or traceroute for Linux and Unix), they notice that they can see just the GGSN's IP address, and the devices after the GGSN only if the operator allows. Direct access to core network equipment is not allowed by nature of mobile cellular networks. The traffic from a subscriber to the Internet, and from the Internet to a subscriber, could affect core network devices dramatically. It is a serious threat and a countermeasure needs to be taken.

The most practical way to decrease the risks here to an acceptable level, without monetary investment, is to enable the IP filter feature of the used Internet router without a firewall feature. With this feature, a router examines the source and/or destination IP address fields, and possibly also the TCP/UDP port numbers, in a packet's IP header. By controlling the TCP/UDP port number in the IP header, it might achieve more detailed filtering. On the other hand, Access Control Lists (ACLs) consist of IP/TCP/ UDP based filtering entries, and they are typically used to indicate the allowed/blocked IP addresses whose traffic is allowed/blocked. This simple countermeasure will protect the core network devices and also the UEs. However, the basic ACLs on the router supports only basic IP/TCP/UDP restriction and does not provide session state tracking or stateful inspection for all the used protocols such as TCP and UDP. Without a stateful firewall feature set, a router can only support TCP SYN attack protection. However, all other attacks with TCP flags, such as ACK, RST/ACK, FIN, FIN/ACK, etc., which are set to 1, can pass through a router's ACLs without being blocked. Therefore, a router recognizes any packet coming from the Internet with its SYN bit set to 1. If the packet is not given the access right on the ACL, it will be dropped. Thus, a router can block only the unwanted packets with their TCP SYN bit set to 1. And all other TCP or UDP packets will be allowed in order not to block the response to legal customer initiated traffic. We call this kind of an ACL as TCP ACL. Hence, by using this kind of a solution, we protect customers and mobile core network infrastructure against only TCP SYN based attacks. Although our routers provide only TCP SYN based attack protection, we notice that the number of paging on RNC decreases after the TCP based Access Control List (ACL) is applied. To ensure the positive effect, we temporarily disabled TCP based ACL on our Internet router and observed the paging activity and resource usage of our core network devices. After we applied TCP based ACL on our Internet routers again, the paging activity and CPU usage was influenced positively as shown with the graphs in Figures. In these graphs, the horizontal axis shows time and date, and the vertical axis shows the CPU usage. These graphs have a sampling period of 1875 s. As seen in these graphs, the CPU usage on the RNC and SGSN decrease by approximately 25%, compared to the values when the ACL is not applied.

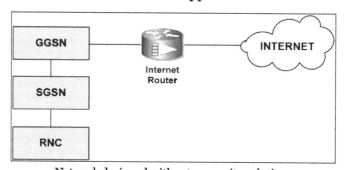

Network designed without a security solution.

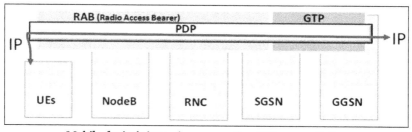

Mobile device's intrusion prevention (IP) connectivity.

These results do not mean that we are under a TCP SYN based attack all the time but there is some TCP based traffic from the Internet to our 3G subscribers' subnet. Therefore, the CPU usage is decreasing after applying the TCP based ACL that blocks the unnecessary traffic from the Internet. Paging activity on the RNC also decreased from 240,000 to 160,000, i.e., by approximately 33%, after applying the ACL on the Internet router, as shown in figure where the vertical and horizontal axes show the number of paging activity and date/time, respectively.

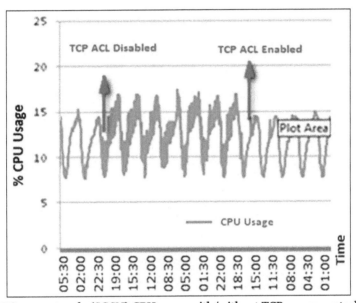

Serving GPRS support node (SGSN) CPU usage with/without TCP access control list (ACL).

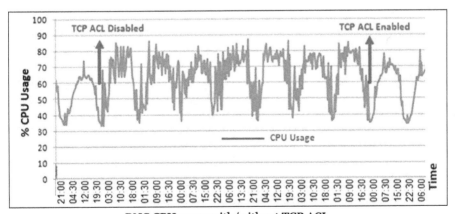

RNC CPU usage with/without TCP ACL.

These results made us also think about the risks that could occur due to UDP based attacks. Therefore, using UDP packets, we performed some attacks against our 3G network. By using some free open source tools, we sent UDP packets of size 60 bytes at a rate of 60,000 to 100,000 packets per second from the Internet to our 3G network. We performed our attacks at midnight between 02:30 and 03:00 a.m. to minimize the adverse effects to the customers. During these attacks, the number of paging and amount of CPU usage on our core network devices dramatically increased as shown in figure. As seen in the figure, the RNC's PS paging attempt count was 70,000 just before attack and during the attack it reached 283,000. Hence, the number of pagings increased by approximately 304% due to our attack.

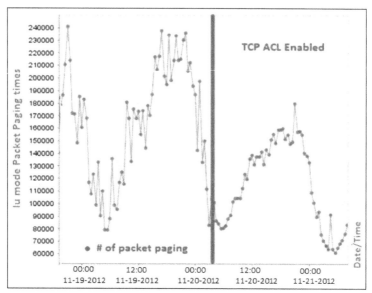

Number of paging activity on an radio network controller (RNC) with/without applying an ACL.

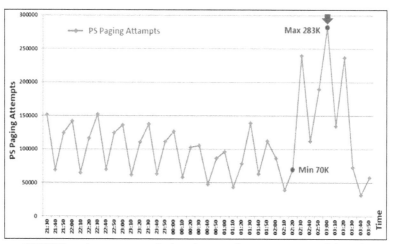

RNC PS paging attempts during the user datagram protocol (UDP) based attacks (sampling period is 30 min).

PS paging activity dramatically increased on the RNC and SGSN when we sent bulk data traffic to the subnets of the 3G subscribers. However, CS paging is not affected during that time. Both the PS and CS paging activity decreased after midnight due to the decreasing number of phone calls. Figure shows how the same RNC's CS paging activity changes during the same time interval. This shows that the PS paging and the CS paging are used for different purposes. The CS and PS paging activities are triggered by the MSC and GGSN, respectively. In other words, PS paging is data-oriented and CS paging is voice-oriented. They use different signaling channels until the RNC. However, they use the same control channel from the RNC to NodeB. Therefore, all voice and data transmission can be effected if it is overloaded by an attacker who exploits PS paging.

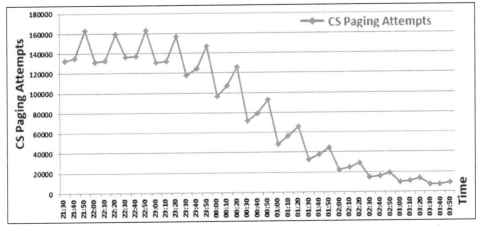

Circuit switched (CS) paging attempts on the RNC during the UDP based attack.

During our attacks, we observed that the CPU usage of an RNC increased. In figure, the RNC CPU usage during the UDP attack is given with a sampling period of 1875 s.

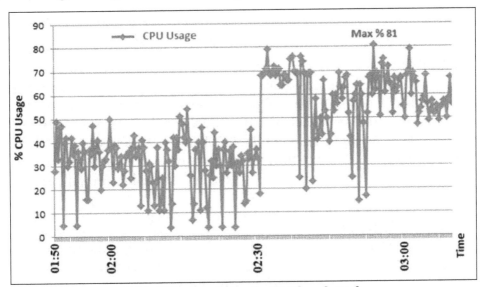

RNCs CPU usage during the UDP based attack.

In this work, we first examined the effect of unwanted data traffic on cellular network devices such as the RNC and SGSN. We realized that blocking unnecessary TCP traffic from the Internet to 3G subscribers' IP subnets decreases not only the resource usage on cellular network devices but also the number of paging attempts on the RNC. However, we did not apply the UDP based access restriction on our Internet router due to the router's limited filtering capability. In order to see the effects of UDP based DoS/DDoS or flooding attacks from the Internet, we performed a UDP based attack at midnight because of the smallest data and voice load on cellular network at that time. The CPU usage and number of PS paging attempts of the RNC was dramatically affected, although the attack was not sophisticated and it was not performed during peak hours. All these results show that mobile cellular networks are more vulnerable against unwanted bulk traffic caused by DoS/DDoS or flooding attacks.

The RNC is one of the most critical device in a mobile network. It is related to both data and voice services. Hence, both voice and data subscribers will be affected if the RNC is out of service due to a CPU overload caused by high traffic resulting from attacks or malicious software such as viruses, worms, etc.

Countermeasures against 3/4G Security Threats

The mobile communication industry has been getting a rise during the last three years. Hence, the security issues in mobile communications have become increasingly more important in 2013. Therefore, not only mobile operators but also mobile device users should be careful when using services over a mobile network. We focused especially on the mobile operators' security vulnerabilities when they don't protect their customers and networks against cyber threats. So, we will mention only the countermeasures that could be taken on the mobile operator's side. Security countermeasures against the security threats on the 3G border and core network devices could be summarized as follows:

- Stateful IP packet filtering could be applied on the Internet border of the mobile network. This task could be achieved by an Internet router with firewall feature with both TCP and UDP stateful packet filtering functionality. A second option is to use a firewall device between the GGSN and the Internet router. The packet filtering device must be stateful for TCP and UDP, and drop all unnecessary traffic from the Internet to the mobile operator or to the customers. The drawback of the stateful firewall solution is the limitations in a firewall's session state table;

- Carrier-grade Network Address Translation (NAT,CGN) is used both to mitigate IPv4 exhaustion and to block unnecessary traffic from the Internet. But when a CGN is used, it is not possible to allow the traffic from the Internet to the subscribers, which is a drawback in using the CGN;

- Configuration should be done against IP address spoofing on the IP packet filter device or the Internet router;

- The flooding and the DoS/DDoS type of attacks are the most dangerous and effective ways to destroy target devices. They could be performed individually or organized, e.g., using a botnet. A high number of connection requests could put most stateful firewalls out of service due to limitations in their state tables. Therefore, a DoS/DDoS protection service or product should be used against attacks coming from the Internet. Generally, it is recommended that DDoS protection is placed in the ISP's side;

- Mobile to mobile traffic could cause spreading of worms and viruses. This can be dangerous for both mobile devices and mobile operators, especially when worms/viruses act in an organized manner forming a botnet. Therefore, mobile to mobile traffic should be filtered by the GGSN or an IP filtering device;

- Malware or botnet members can generate flooding traffic to their victims such as mobile devices or operators, causing service interruptions on core network devices. Hence, rate limiting should be applied to mobile user traffic on the GGSN or an IP packet filtering device should be used to avoid this threat;

- Directed broadcast traffic must be denied on the GGSN to protect against DDoS attacks;

- RA and LA scope optimization should be done according to the best practices to decrease paging effects;

- Parameter optimization on RNC and SGSN should be done according to the best practices to decrease the PS paging effect;

- Mobile operators could locate the GTP aware firewall products between the SGSN and the GGSN to block attacks coming through the GTP protocol;

- The PS and the CS paging control channels should be isolated to prevent the CS paging from being overloaded by the PS paging channel under attack. Hence, at least voice communication service would not be affected in the case of an attack against the PS paging channel.

A pictorial summary of all the countermeasures to be taken on the mobile operator's side can be seen in figure. The advantages and disadvantages of using the above mentioned countermeasures are listed in the table.

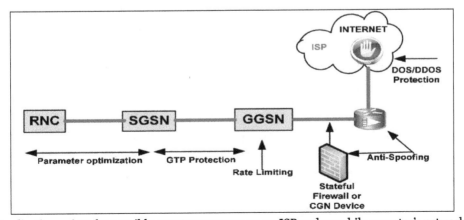

Application points for possible countermeasures on an ISP and a mobile operator's network.

Advantages and disadvantages of possible countermeasures against DoS and DDoS attacks on 3G mobile networks.

Protections	PROS	CONS
DDoS Protection on ISP	Effective DDoS protection.	Extra cost.
Stateful Firewall	Effective and flexible IP filtering.	Session state table limitations, Point of failure.
CGN	Blocking unwanted traffic from the Internet.	NAT table limitations, Lack of IP filtering, Point of failure.
GTP Firewall	GTP attack protection.	Extra cost, Point of failure.
Parameter Optimization	No-cost and protection.	N/A
User Isolation	Preventing organized internal attacks.	Unforeseen UE-to-UE communication.
Rete Limiting	Decreases the effect of attacks.	N/A

Security solutions could be applied most easily and efficiently on the border of the Internet. We have shown that cellular mobile networks should be effectively protected by using firewalls, Intrusion Prevention Devices (IPS) and/or DoS/DDoS protection products on the border of the Internet. In addition to these solutions, DoS/DDoS protection on the ISP side is very important, due to the limitations in the session tables of stateful firewalls. DDoS protection service on the ISP side is expected to filter all flooding and DDoS attacks before they overload an Internet connection link and fill up the session table of a stateful firewall. Another security solution is to isolate the customer traffic in order to prevent spreading of malicious software which may otherwise result in subscriber dissatisfaction due to additional usage charges and also harm core network equipment.

Requirements for Communication Security

Communication security is often described in terms of confidentiality, integrity, authentication and nonrepudiation of transmitted data. These security services are in turn implemented by various mechanisms that are usually cryptographic in nature. (ISO 1988) for a concise description of communication security services and mechanisms. In addition there is confidentiality of traffic (i.e. whether or not communication is taking place), of location (where the communicating parties are located) and of the communicating parties' address, all of which are important for privacy. A casual level of security is usually provided implicitly even without taking any extra measures. For example in order to eavesdrop on a particular person's mobile phone conversations the eavesdropper has to be located in physical proximity to the person and carry special radio equipment which in itself represents a certain level of protection. Casual authentication between mobile phone users is indirectly provided by the calling and called party numbers. In case of of voice telephony, authentication results from recognising the other person's voice.

Cryptography on the other hand gives the possibility of designing strong security services but often creates inconveniences when using the application. The use of cryptography therefore makes most sense in case of sensitive applications. When strong cryptographic security mechanisms are in place the remaining vulnerabilities are usually due to poor management and operation and not by weaknesses in the cryptographic algorithms themselves.

Confidentiality of transmitted data can be provided by encrypting the information flow between the communicating parties, and the encryption can take place end-toend between the communicating parties or alternatively on separate legs in the communication path. In GSM networks for example, only the radio link between the mobile terminal and the base station is encrypted whereas the rest of the network transmits data in clear-text. Radio link confidentiality in GSM is totally transparent from the user's point of view. Mechanisms for implementing confidentiality of traffic, location and addresses will depend on the technology used in a particular mobile network.

Authentication of transmitted data is an asymmetric service, meaning for example that when and are communicating, the authentication of 's data by is independent from the authentication of 's data by . The types of authentication available will depend on the security protocol used. In the Internet for example, SSL allows encryption with four different authentication options: 1) server authentication, 2) client authentication, or 3) both server and client authentication or 4) no authentication, i.e. providing confidentiality only.

Non-repudiation is similar to authentication in that it is an asymmetric security service. A simple way to describe the difference between authentication and non-repudiation is that with authentication the recipient himself is confident about the origin of a message but would not necessarily be able to convince anybody else about it, whereas for non-repudiation the recipient is also able to convince third parties. Digital signature is the mechanism used for non-repudiation. Cryptographicly seen a message's authentication code and non-repudiation code can be identical, and the difference between the two services might only depend on the key distribution. In general, if a signature verification key has been certified by a trusted third party the corresponding digital signature will provide nonrepudiation, whereas it can only provide authentication if the key has simply been exchanged between the two communicating parties.

Different parties will have different interests regarding authentication and non-repudiation services. Network operators are interested in authenticating the users for billing purposes and to avoid fraud. Users and content service providers are interested in authenticating each other and might also be interested in authenticating the network service provider. How and where in the network authentication services are implemented will depend on the technology used and the business models involved.

The Network Operator as Trusted Third Party

Public-key cryptography is the basis of several important security services such as non-repudiation and authentication and is an essential element for SSL that is used for securing Web communication. One public/private key pair is used for authenticating one party by the other, and mutual authentication requires two key pairs. In fact, every entity on the Internet needs a key pair if it shall be possible for an arbitrary entity to authenticate any other entity. It has therefore been predicted that every player on the Internet will have its own public/private key pair which will form the basis for the user's or organisation's digital identity in electronic environments. This requires the secure generation and distribution of potentially hundreds of millions of public/private key pairs, which poses a formidable key management challenge.

A PKI refers to an infrastructure for distributing public keys where the authenticity of public keys is certified by Certification Authorities (CA). A public key certificate basically consists of the CA's digital signature on the public.

key, usually together with some attributes. If the certificate owner's identity is one of the attributes, then the certificate is called an identity certificate, and the purpose of the certificate is to link the public key and the identity together in an unambiguous way. The CA is a Trusted Third Party (TTP) because it is trusted to correctly verify and certify the identity of the public-key owner before issuing the certificate. The structure of identity public-key certificates is standardised by the ITU X.509 standard (ITU 1997). In order to verify a certificate the CA's public key is needed, thereby creating an identical authentication problem. The CA's public key can be certified by another CA etc., but in the end you need to receive the public key of some CA, usually called the root CA, out-of-band in a secure way. This is difficult to achieve with a handful of global CAs serving the whole Internet community as the case is for the Web PKI. If CAs are either local and/or serve a limited number of relying parties then trust relationships can be much stronger, and out-of-band distribution of CA root public keys and user private keys can be much more secure.

In case of subscription based mobile networks there exists a formal relationship between users/subscribers on one hand and the network operator on the other. It would therefore be natural to let the network operator play the role of CA. The user's private key as well as the root CA public key can be distributed in a secure way based on the distribution of subscription tokens e.g in the form of the GSM SIM card. Operators who have formed roaming agreements between each other already have a formal relationship which could be extended to cross-certification of each other public keys. Mobile network operators therefore are in a very strong position to establish themselves as CAs, and the mobile device, or more precisely the security token, naturally lends itself to become a secure storage medium for these cryptographic keys. The Web PKI suffers from insecure distribution and storage of cryptographic keys and therefore does not provide a complete chain of trust. To combine the roles of CA and mobile network operator would make it easier to have a complete chain of trust around the PKI because there already exists a trust relationship between mobile network operators and their customer.

Network operators should also explore the possibility of becoming close partners with financial institutions or alternatively establish themselves as independent financial mediators by allowing m-commerce transactions to be billed on subscriptions. This has already happened on a small scale in cases when customers can buy e.g. soft drinks from vending machines by placing a call to a premium rate number linked to the vending machine. An evolution from this primitive type of payment to a more general and flexible form could be driven by the network operators. One of the major problems in e-commerce is the lack of customer authentication. The fact that network operators already have strong subscriber authentication puts them in a natural position to become an intermediary between customers and vendors. This will require a relationship between vendors and network operators similar to the relationship between vendors and credit card companies.

Security Across Heterogeneous Network

Network architectures are based on protocol layers which represent an abstract way of modelling and implementing data transmission between communicating parties. The usual protocol architecture consists of 5 layers as illustrated in Figure below.

In reality, no data are directly transferred between adjacent layers on opposite sides. Instead, data and control information are passed down through the interfaces between the protocol layers on one side and up through the interfaces between the protocol layers on the other side. The physical data transmission actually takes place through a physical medium underneath the physical protocol layer.

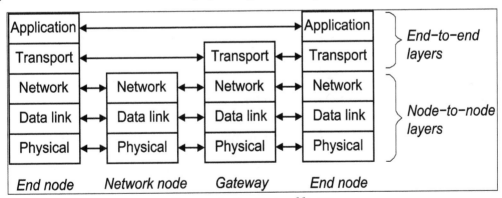

Communication protocol layers.

The physical, data-link and network layers are node-tonode whereas the transport and application layers usually are end-to-end except when there is a gateway in between. Figure shows an example with a gateway in the transport layer in which case the transport layer no longer can be considered end-to-end.

The network architecture and the security goal together indicate the most appropriate protocol layer where a security service is to be located. Authentication and nonrepudiation are for example only meaningful when implemented end-to-end between the parties that need to authenticate each other. Confidentiality and integrity on the other hand can be meaningful by encrypting isolated legs between nodes when it can be assumed that only these legs would be vulnerable to attack. By using Figure as an example, authentication and non-repudiation must thus be implemented on the application layer, whereas confidentiality can be implemented on any layer.

Mobile applications usually span over several networks such as for example a radio network and a fixed network requiring gateways on the transport or application protocol layers. This complicates the implementation of security services because it becomes more difficult to obtain end-to-end security. As a general rule authentication must always be built on top of an end-to-end layer. Whenever confidentiality is based on encryption with a session key obtained through the authentication protocol it is natural to let encryption be end-to-end as well.

An example of a mismatch between desired security service and protocol layer can be seen in the original WTLS protocol (Wireless Transport Layer Security) (WAP-Forum 2000a). WTLS is intended to work similarly to SSL for example by providing authentication. However because the WTLS protocol terminates in the transport layer gateway it is not able to provide authentication between the WAP terminal and the WAP service provider, but only between the WAP terminal and the WAP gateway as illustrated in figure below.

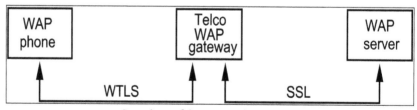

Security architecture using WTLS.

In addition to making authentication meaningless, this solution also creates an unavoidable plain text gap in the WAP gateway. WTLS is specified with 3 functionality classes so that the security features can be introduced in steps. Table describes the functionality of each class, where "M" means mandatory and "O" means optional.

Class 1 Class 1 is being used in specifies public-key exchange without server or client certificates and is based on the Diffie-Hellman key exchange protocol, which provides encryption and confidentiality but no authentication. So far only WTLS mobile WAP applications. Class 2 requires using server certificates and is supposed to provide server authentication. Class 3 requires using client certificates and is supposed to allow user authentication by the service providers.

However, when considering that the original WTLS architecture does not provide end-to-end security it is obvious that the WTLS classes 2 and 3 would be meaningless. The original WTLS can only ever provide authentication of the WAP gateway which is of no value to users or WAP service providers. What users and service providers want is to be able to authenticate each other.

WTLS Classes

Feature	Class 1	Class 2	Class 3
Public-key-exchange	M	M	M
Server certificates	O	M	M
Client certificates	O	O	M
Shared-secret handshake	O	O	O
Compression	-	O	O
Encryption	M	M	M
MAC	M	M	M
Smart card interface	-	O	O

We find it surprising that WTLS originally suffered from this serious design weakness considering that the development of this technology was given top priority by some of the world's most prestigious IT and telecommunications companies. One possible explanation for introducing the WAP gateway could have been to give the mobile network operators more control of the traffic and transactions and thereby allow specific business models. It could also have been to facilitate legal government interception of traffic contents from the WAP gateway clear text gap. With an end-to-end security architecture this would change, and in fact become similar to the existing security architecture on the Internet.

Because of the deficiencies in the original WTLS protocol the WAP forum has defined a new standard for endto-end SSL using tunnelling through the WAP gateway (WAP-Forum 2000b). This is achieved by implementing a wireless enhanced version of the Internet TCP transport protocol layer in the mobile devices and run SSL on top of that. However, because the origin server will probably not support wireless enhanced TCP, there will be a proxy that acts as the termination point of two TCP sessions, one w-TCP to the client mobile terminal and one TCP to the server. It will just move packets across transparently from one connection to the other.

Usability of Security

Details of the security services and mechanisms are often complex and users would quickly be overloaded with information if the details were presented to them. A common design philosophy is therefore to make security services and mechanisms as transparent as possible. However there is a danger that users receive too little security information. If security is totally hidden from the user he or she would not be able to tell whether it is working the way it was intended, which in turn could allow successful attacks to remain undetected. Obviously, the security evidence provided can not be more than the user can understand and handle but it must be sufficient for the required security level of the application. The challenge is to determine what type of evidence is really necessary and present it to the user in an intuitive and intelligible way.

In the computer network jargon, it is sometimes forgotten that communication ultimately goes between human users and organisations, and that some security services only are meaningful if they are designed to suit human users. The interpretation of communication in the human brain can conceptually be described as a semantic protocol layer above the application layer as depicted in figure below.

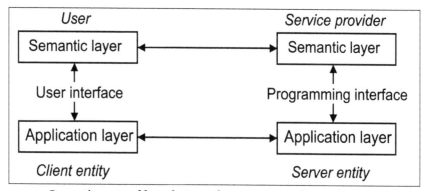

Semantic protocol layer between human users and organisations.

Between the application and the semantic layers lies the user interface, and this is not something that can be specified and implemented in a software module as it is done for the other protocol layer interfaces. A good user interface represents an intricate combination of multimedia and optimal terminal design adapted to the human physical and mental capabilities. The study of how security information should be handled in the user interface forms part of security usability. This field of research has mostly been ignored by researchers as well as application and hardware developers. Whitten and Tygar argue that effective security requires a different usability standard, and that it can not be achieved through the user interface techniques appropriate to other types of consumer software.

One example illustrating the subtleties of security is the padlock icon on Web browsers where an open padlock indicates insecure communication whereas a closed padlock indicates secure communication. This is seemingly a very neat and intuitive way of indicating that a Web server has been authenticated with SSL and that transmitted and received data are being encrypted. However a closed padlock only tells the user that some Web server has been authenticated but not which Web server in particular. As long as the user does not do the extra mouse clicks to view the server certificate he or she has not authenticated anything at all. Despite the appearance of being very simple, the padlock hides crucial aspects of security, without which authentication becomes meaningless.

The Web browser does allow viewing the server public-key certificate by clicking on the padlock icon, but users hardly ever do this, and even security aware users who view the certificate when accessing a secure Web site can have difficulty in judging whether the certificate really is what it claims to be. The browser usually checks that the domain name in the certificate is the same as the domain name pointed to by the browser, and aware users might notice when an intruder's domain name is different from the expected domain name. However, users do not usually inspect the URL for the domain name when browsing the Internet Web, and many companies' secure Web sites have URLs with non-obvious domain names. that do not correspond to the domain names of their nonsecure Web sites. One example is the Norwegian bank Nordea with the URL: `http://www.nordea.no` and where its secure on-line banking has URL: `https://ibank.bbsas.no/iBank/Dispatcher`. Another vulnerability is the fact that distinct domain names can appear very similar, for example differing only by a single letter so that a false domain name may pass undetected. How easy is it for example to distinguish between the following URLs: `http://www.bellabs.com`, `http://www.belllabs.com`, and `http://www.bell-labs.com` ?

In order to make authentication on the Internet Web more meaningful, some familiar elements from the physical world could be used. Jøsang et al. have proposed to display a digitally certified company logo in the Web browser to allow meaningful authentication at a glance and bridge the gap between the cryptographic mechanisms and the human user. This idea is presently discussed in the IETF and may become a standard feature in the future.

For mobile devices, the relatively small visual display will make it virtually impossible to inspect public-key certificates for authentication. Cryptographic authentication by identity certificates such as X.509 will be unreliable because of the difficulty of comparing an Internet site name with the identity stored in the digital certificate. Figure shows a typical hand-held WAP device with which server authentication will be meaningless.

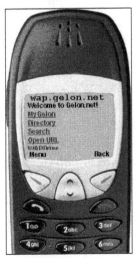

Interface that is unable to provide meaningful WAP server authentication.

By typing the correct URL of a WAP or Web site, authentication is not really needed as long as the integrity of the network is preserved, i.e. you will access the right site as long as you type the URL correctly. WAP sites are more likely to be accessed through portals than by typing URLs, which makes other forms of authentication the more important. However, cryptographic authentication mechanisms are only meaningful if the interface is able to provide authentication information in a secure way. For mobile devices with small display, certified company logos seem to provide a good solution.

The integrity of the evidence presented to the user can be assured by having a reserved area for certified content on the interface which is never used for other types of content. Because of limited size of visual displays this might seem to be an expensive sacrifice. We therefore recommend using the normal display for displaying security information, but in a special security mode, and instead to reserve a small exclusive area to indicate that the display is in security mode. The exclusive security display area and the security display mode should not be accessible by content applications. This security mode should be easy to invoke and be distinguishable from the other display modes. The security mode of the interface then represents a separate interface channel that can be distinguished from the normal information content channel.

What represents the most suitable type of certified information to be displayed will depend on the application. A simple solution from an implementation point of view is to link the authentication directly to the logical network address used such as e.g. a telephone number or Internet domain name, and display the certified address in the separate control field. The user would then be required to know exactly which network address he or she wants to contact, but this can be problematic as mentioned above.

Certified company logos, pictures of persons and sound files can be easier to perceive as distinguished qualifiers than simple names. There are however new problems that need to be solved before such solutions can be implemented.

Image and sound can only be used for strong authentication if the image and sound files are certified and included in digital certificates. This requires the CA to verify their authenticity before issuing certificates. A company logo must for example be sufficiently different from all other company logos and this requires the CA to perform a similarity check, but this is likely to create new problems. What are the criteria for a similarity check? If similar logos or names are used by companies in totally different businesses, is that OK? According to Stubblebine and Syverson hierarchies adequate to issue certificates are not by themselves adequate to ensure global uniqueness. Suppose that a company obtains a certificate for a logo and then another company applies for a certificate for a much too similar logo, but it owns that logo as a registered trademark? More generally, what about revocation of a logo because of previously unrecognised problems? Does every little shop need to hire a graphic artist? What is the size of the space of meaningfully discernible logos? The authenticity of pictures of persons can best be assured by taking the photos on the CA's premises. Similar requirements apply to sound files, i.e. they must be recorded in person on the CA's premises.

Verifying these additional elements will require the CA's to be physically more local to users or organisations, making it difficult for one CA to serve the whole world. In Sec.3 we described how PKI operations can be simplified by letting the mobile network carrier act as CA.

Securing Active Contents

Before active content was available, Web pages were mainly static displays of information coded in the Hyper Text Markup Language (HTML). Active content allows sound and image animation and provides the user with the ability to interact with the server side during a Web session. Active content exists in many forms. Java applets and ActiveX controls are some of the best known but there are also JavaScripts, VBScripts, MSWord Macros and even images. All these basically consist of mobile code that is sent from the Web server and loaded into the client machine for execution there.

All this is very appealing from a functionality and flexibility point of view but it poses a formidable threat to the integrity of the client machine. Active content can cause damage by intent or by simply being poorly designed. A discussion of threats and risks posed by active contents can be found in. An attack using malicious applets is described in. Firewalls offer little protection because they are usually configured to let http traffic and active content through. Unless the active content can be controlled, all files and network connections can be accessed and (mis)used, making it impossible to operate any secure applications on the client machine. Sandboxing and certification can be used to counter threats from active content.

Sandboxing basically means that the active content is constrained in what resourcesit can access on the host system. The advantage is that it is always active and completely transparent to the user. The disadvantage is that it severely limits the capabilities of active contents.

Certification means that a trusted party has validated and digitally signed the active content and that the platform verifies the digital signature before it can execute. The advantage is that the active content can access all system resources. The disadvantage is that certification is not equivalent with trustworthiness. A Web browser can for example be tuned so that any piece of certified active content is accepted by default or alternatively so that only active content certified by certain parties is accepted by default and that any other trigger a dialog box. The dialog box basically asks the user whether he or she wants the active content to be executed. Experience shows that users almost always accept active content when asked by a dialog box simply because they want the functionality and because most active content is benign anyway. This means that should the user receive a piece of malicious active content he or she will almost certainly make the wrong decision and accept it. The user simply does not have sufficient evidence to make an informed decision.

A similar development is taking place in the market for downloadable executables in mobile terminals, but the maturity of this technology is still behind that of the Web, and due to technical constraints is likely to follow a slightly different path.

Presently there is no standard protocol for downloading executables to mobile terminals such as http on the Web. Neither is there a standard execution environment for running executables on mobile terminals such as the Java Virtual Machine in Web browsers. WAP was originally perceived as a method by which all types of content would be downloaded, including WMLScripts which allow minimally executable applicationsto be run on mobile terminals. However, WAP never achieved its envisioned acceptance in the market, and was not designed to provide an execution environment for programs written in Java or other rich programming languages.

Several stakeholders have started to roll out new technology to correct these deficiencies, and the question to ask is whether these solutions will provide the necessary security to protect the mobile terminals against harmful active contents.

Sun Microsystems has introduced Java 2 MicroEdition (J2ME) and the Kilobytes Virtual Machine (KVM) to provide an application execution environment for constrained devices such as mobile terminals. A complementary and compatible technology for downloading content and executables to mobile terminals called "Download Fun" has been introduced by Openwave Systems. According to the Download Fun FAQ it provides a mechanism to download binary objects from a content site to a mobile device in a secure manner. What that really means is that the Download Fun Client supports a variety of security protocols including SSL and WTLS Class II, meaning for example that the executables can be digitally signed, and that the signature must be verified before the executable can run on the mobile terminal. The network operator will normally sign the executable, and a revenue sharing scheme will make sure that third party application developers get included in the revenue stream.

Qualcomm has also entered the market with Binary Runtime Environment for Wireless (BREW). BREW encompasses both an application execution environment and a mechanisms for downloading executables. Qualcomm has also introduced a scheme for digitally signing executables, and this forms part of BREW's particular business model. Qualcomm will be the only authority to validate BREW applications, and the digital signature will be applied by Qualcomm, the application developer and the network operator in concert. In that way Qualcomm makes sure it always gets included in the revenue stream. The reason why Qualcomm believesthe market will accept this type of monopolistic business model is that they control the production of chips for CDMA1 phones, and will make sure that every CDMA chip is BREW-enabled at no extra cost to the mobile phone manufacturers.

What remains to see is whether these two (and other) technologies will restrict users to only download and run digitally signed executables. That would require the mobile terminals to be designed so that the network operator controls what executables the terminals can run. Our guess is that users will want solutions that give greater flexibility and allow running any executable if they so desire, which necessarily will create security vulnerabilities. In fact mobile phones have already been the target of various types of malicious active content as for example reported in.

As a result, additional security mechanisms are needed to protect against harmful executable. Using sandboxing and dialog boxes obviously comes to mind, but unfortunately these mechanisms seem even less suitable in mobile terminals than they are in Web browsers. The challenge is therefore to come up with alternative and better solutions. Present mobile phones do not have enough memory to run traditional anti-virus software. A solution suggested by Hoshizawa is to run anti-virus software at the network level so that network operators and ISPs can block virus outbreaks and thereby prevent them from spreading.

On a non-technical level, it is of course always a good idea to improve user awareness and hygiene in download habits. This may seem like daunting task given that it would require educating the global mass consumer market. Nevertheless network operators should have an obligation to make an effort towards greater awareness about mobile phone security.

GSM Security

Security Mechanisms

GSM has a lot of security systems to build safe communication. It includes a lot of different types of algorithms and different type of devices.

The main security measurements of GSM security can be written in 4 principles;

Authentication of a user; it provides the ability for mobile equipment to prove that it has access to a particular account with the operator.

Ciphering of the data and signaling; it requires that all signaling and user data (such as text messages and speech) are protected against interception by means of ciphering.

Confidentiality of a user identity; it provides IMSI's (international mobile subscriber identity) security. GSM communication uses IMSI rarely, it uses TMSI (Temporary Mobile Subscriber Identity) to provide more secure communication and to avoid disclosing of user's identity. This means someone intercepting communications should not be able to learn if a particular mobile user is in the area.

Using SIM as security module; Incase SIM card was taken by opponent, there is still PIN code measurement.

A3 and A8 Algorithms

A3 and A8 algorithms are A3 and A8 algorithms are symmetric algorithms which the encryption and decryption use the same key. Both of the algorithms are one way function, it means that output can be found if the inputs are known but it is mostly impossible to find inputs incase the output is known. A3 and A8 algorithms are kept and implemented in SIM card.

Many users of GSM will be familiar with the SIM (Subscriber Identity Module) the small smartcard which is inserted into a GSM phone.

Sample SIM Card.

The SIM itself is protected by an optional PIN code. The PIN is entered on the phone's keypad, and passed to the SIM for verification. If the code does not match with the PIN stored by the SIM, the SIM informs the user that code was invalid, and refuses to perform authentication functions until the correct PIN is entered. To further enhance security, the SIM normally "locks out" the PIN after a number of invalid attempts (normally 3). After this, a PUK (PIN Unlock) code is required to be entered, which must be obtained from the operator. If the PUK is entered incorrectly a number of times (normally 10), the SIM refuses local access to privileged information (and authentication functions) permanently, rendering the SIM useless.

Typical SIM features can be lined as below:

- 8 bit CPU.

- 16 K ROM.

- 256 bytes RAM.

- 4K EEPROM.

- Cost: $5-50.

On its own, the phone has no association with any particular network. The appropriate account with a network is selected by inserting the SIM into the phone. Therefore the SIM card contains all of the details necessary to obtain access to a particular account. It contains 4 important information; IMSI, Ki, A3 and A8 algorithms.

IMSI (International Mobile Subscriber Identity): Unique number for every subscriber in the world. It includes information about the home network of the subscriber and the country of issue. This information can be read from the SIM provided there is local access to the SIM (normally protected by a simple PIN code). The IMSI is a sequence of up to 15 decimal digits, the first 5 or 6 of which specify the network and country.

Ki: Root encryption key. This is a randomly generated 128-bit number allocated to a particular subscriber that seeds the generation of all keys and challenges used in the GSM system. The Ki is highly protected, and is only known in the SIM and the network's AuC (Authentication Centre). The phone itself never learns of the Ki, and simply feeds the SIM the information it needs to know to perform the authentication or generate ciphering keys. Authentication and key generation is performed in the SIM, which is possible because the SIM is an intelligent device with a microprocessor.

A3 Algorithm: It provides authentication to the user that it has privilege to access the system. The network authenticates the subscriber through the use of a challenge-response method.

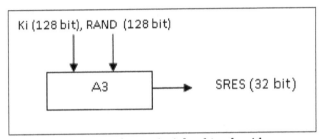

Illustrates working principle of A3 algorithm.

Firstly, a 128 bit random number (RAND) is transmitted to the mobile station over the air interface. The RAND is passed to the SIM card, where it is sent through the A3 authentication algorithm together with the KI. The output of the A3 algorithm, the signed response (SRES) is transmitted via the air interface from the mobile station back to the network. On the network, the AuC compares its value of SRES with the value of SRES it has received from the mobile station. If the two values of SRES match, authentication is successful and the subscriber joins the network. The AuC actually doesn't store a copy of SRES but queries the HLR or the VLR for it, as needed.

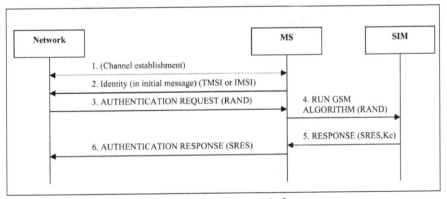

A3 Algorithm Request Order.

Figure shows the request order between mobile station and operator network in A3 algorithm. This figure can be explained as:

- Some connection is attempted between the phone and the network.

- The phone submits its identity. All potential messages used at the start of a connection contain an identity field. Where possible, it avoids sending its IMSI in plaintext (to prevent eavesdroppers knowing the particular subscriber is attempting a connection). Instead, it uses its TMSI (Temporary Mobile Subscriber Identity).

- The network sends the AUTHENTICATION REQUEST message containing the RAND.

- The phone receives the RAND, and passes it to the SIM, in the RUN GSM ALGORITHM command.

- The SIM runs the A3 algorithm, and returns the SRES to the phone.

- The phone transmits the SRES to the network in the AUTHENTICATION RESPONSE message.

- The network compares the SRES with its own SRES. If they match, the transaction may proceed. Otherwise, the network either decides to repeat the authentication procedure with IMSI if the TMSI was used, or returns an AUTHENTICATION REJECT message.

A8 Algorithm: GSM makes use of a ciphering key to protect both user data and signaling on the vulnerable air interface. Once the user is authenticated, the RAND (delivered from the network) together with the Ki (from the SIM) is sent through the A8 ciphering key generating algorithm, to produce a ciphering key (Kc). The A8 algorithm is stored on the SIM card. The Kc created by the A8 algorithm, is then used with the A5 ciphering algorithm to encipher or decipher the data. The A5 algorithm is implemented in the hardware of the mobile phone, as it has to encrypt and decrypt data on the air.

Whenever the A3 algorithm runs to generate SRES, the A8 algorithm is run as well The A8 algorithm uses the RAND and Ki as input to generate a 64-bit ciphering key, the Kc, which is then stored in the SIM and readable by the phone. The network also generates the Kc and distributes it to the base station (BTS) handling the connection.

A8 algorithm working principle:

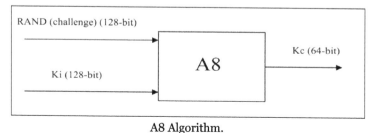

A8 Algorithm.

COMP128

COMP128 is hash function which is an implementation of the A3 and A8 algorithms in the GSM standard.

The Algorithm Expert Group was held in 1987 and designed GSM encryption algorithms. They created 2 algorithms; the first one was COMP128 which is to provide authentication and derive the cipher key (A3/8), the second was A5 algorithm. GSM allowed every operator to use its own A3/8 algorithm and the all system support this without transferring between networks, also during roaming. However, most of the operators do not have expertise to make their own A3/8 algorithm design and they use example COMP128 design.

The COMP128 takes the RAND and the Ki as input; it generates 128 bits of output. The first 32 bits of the 128 bits form the SRES response; the last 54 bits of the COMP128 output form the session key, Kc. Note that the key length at this point is 54 bits instead of 64 bits, which is the length of the key given as input to the A5 algorithm. Ten zero-bits are appended to the key generated by the COMP128 algorithm. Thus, the key of 64 bits with the last ten bits zeroed out. This effectively reduces the key space from 64 bits to 54 bits. This is done in all A8 implementations, including those that do not use COMP128 for key generation, and seems to be a deliberate feature of the A8 algorithm implementations.

A5 Algorithm

A5 is a stream cipher which can be implemented very efficiently on hardware. There exist several implementations of this algorithm, the most commonly used ones are A5/0, A5/1 and A5/2 (A5/3 is used in 3G systems). The reason for the different implementations is due to export restrictions of encryption technologies. A5/1 is the strongest version and is used widely in Western Europe and America, while the A5/2 is commonly used in Asia. Countries under UN Sanctions and certain third world countries use the A5/0, which comes with no encryption.

A5/1 algorithm uses the structure which can be seen in figure.

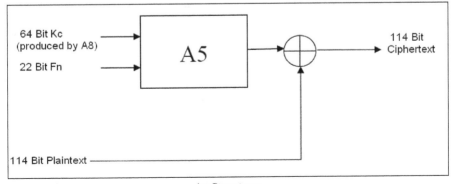

A5 Structure.

As a stream cipher, A5 works on a bit by bit basis (and not on blocks, as DES and AES).

So, an error in the received cipher text will only result in the corresponding plaintext bit being in error.

None of the algorithms are published by GSM Association. They are all discovered by using reverse engineering methods.

Kc is the key which was produced by A8 algorithm. Plaintext is the data which is wanted to transmit. Fn is the frame bits which come from LFSR (Linear Feedback Shift Register) process.

To understand A5/1 algorithm, LFSR (Linear Feedback Shift Register) structure should be introduced firstly.

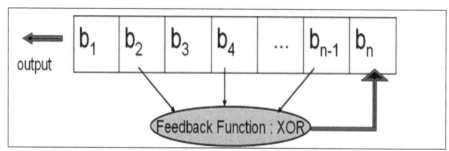

LFSR Structure.

In LFSR structure, there is certain amount of bits which has some special bits (taps). These special taps are XOR ed, all the bits shit one bit left and the result was put to the first bit.

In A5/1 algorithm, LFSR structure uses 3 register bits. These bits are shown in the figure.

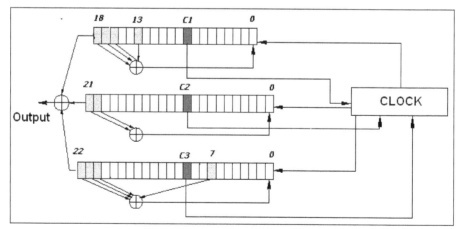

LFSR Structure in A5/1 Algorithm.

A5/1 is built from three short linear feedback shift registers (LFSR) of lengths 19, 22, and 23 bits, which are denoted by R1, R2 and R3 respectively. The rightmost bit in each register is labeled as bit zero. The taps of R1 are at bit positions 13, 16, 17, 18; the taps of R2 are at bit positions 20, 21; and the taps of R3 are at bit positions 7, 20, 21, 22. When a register is clocked, its taps are XOR ed together, and the result is stored in the rightmost bit of the leftshifted register. The three registers are clocked in a stop/go fashion using the following majority rule: Each register has a single "clocking" tap (bit 8 for R1, bit 10 for R2, and bit 10 for for R3); each clock cycle, the majority function of the clocking taps is calculated and only those registers whose clocking taps agree with the majority bit are actually clocked. Note that at each step either two or three registers are clocked, and that each register moves with probability 3/4 and stops with probability 1/4.

The process of generating pseudo random bits from the session key Kc and the frame counter Fn is carried out in four steps:

- The three registers are zeroed, and then clocked for 64 cycles (ignoring the stop/go clock control). During this period each bit of Kc (from least significant bit to most significant bit) is XOR ed in parallel into the lsb's of the three registers.

- The three registers are clocked for 22 additional cycles (ignoring the stop/go clock control). During this period the successive bits of Fn (from lsb to msb) are again XOR'ed in parallel into the lsb's of the three registers. The contents of the three registers at the end of this step are called the initial state of the frame.

- The three registers are clocked for 100 additional clock cycles with the stop/go clock control but without producing any outputs.

- The three registers are clocked for 228 additional clock cycles with the stop/go clock control in order to produce the 228 output bits. At each clock cycle, one output bit is produced as the XOR of the msb's of the three registers. 114 bits of these 228 bits are used in MS-BTS communication and the rest 114 bits are used in BTS-MS communication.

A5/2 is the weak version of A5/1. Different from A5/1, it contains 4 LFSR bits Time complexity of the A5/1 is 2^{54} (if the last 10 bits of the Kc is not zero, then 2^{64}). However, time complexity of A5/2 is 2^{16}. A5/0 is the weakest version between these 3 algorithms. It doesn't make any encryption.

GPRS Security

The security in GPRS is based on the same mechanisms as of GSM. However, GPRS uses a different encryption key to provide security. To authenticate the customer, the same A3/8 algorithms are used with the same Ki, different RAND. The resulting Kc is different than voice communication key and this Kc is used to encrypt GPRS data. This Kc is refered GPRS-Kc to make it different from voice communication Kc. Similarly, SRES and RAND are referred as GPRS-SRES and GPRS-RAND. GPRS cipher is also referred to GPRS A5 or GEA (GPRS Encryption Algorithm).

Weakness of GSM Security

Weak Sides of the Security Mechanisms

GSM doesn't have perfect security system. Opponents can eavesdrop the channel in real time.

First of all, most of the operators do not have expertise enough to create new A3/8 algorithms. So they use COMP128 function without even changing it. This is big security problem because all the COMP128 function has found by reverse engineering.

The bit size of the algorithms is weak. A5/1 algorithm uses 64 bit Kc in the best case. Most operators use COMP128 which has 54 bit Kc and last 10 bits are always zero. Also A5/2 is weaker than A5/1.

Moreover, authentication query only exists BTS-MS communication. There is no authentication for MS-BTS. It means that, fake base stations can behave like real BTS and MS will answer each SRES request from them. The network does not authenticate itself to a phone. This is the most serious fault in GSM security, which allows a man-in-the-middle attack. This weakness was known for GSM constructors at the time of the GSM design, but it was expected that building a false BTS would be too expensive and it would be difficult to make those attacks cost effective. However, after 20 years the situation changed significantly. Today there are companies that product short range BTS, so an attacker can simply buy a BTS at a reasonable price.

Another serious vulnerability of the GSM is the lack of proper Caller ID or Sender ID verification. In other words, the caller number or SMS sender number could be spoofed. The caller ID and the voice is transmitted in different channels. So, Called ID or SMS ID can be spoofed.

Another weakness attackers can exploit is vulnerability in the IMSI protection mechanism. As mentioned before, networks use TMSI to protect IMSI but if the network somehow loses track of a particular TMSI it must then ask the subscriber their IMSI over a radio link. The connection cannot be ciphered because the network does not know the identity of the user, and thus the IMSI is sent in plain text. The attacker can thus check whether a particular user (IMSI) is in the vicinity.

Popular Attacks Types

Capturing One or Several Mobile Stations

In many of the attacks the attacker needs to pretend the network to the MS or pretending the MS to the network or both in a so called man-in-the middle attack. An attacker impersonating BS and MS to each other can eavesdrop, modify, and erase, order, replay, spoof and relaying signals/user data between two communicating entities. The required equipment is an adjusted and modified BTS and MS bundle. The modified BTS behaves as the identity the network to the MS, while the modified MS impersonates the MS to the network.

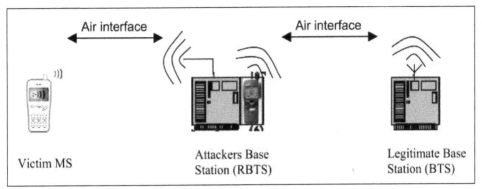

Man-in-Middle mechanism.

Before an active attack, the attacker may have to eavesdrop the MS. The attacker may want to learn the information consists of cell identity, network identity, and control channel structure, list of channels in use and details of the access protocol. In this manner, an attacker with a Fake BTS, providing higher power levels than the BTS, between the victim MS and legitimate BTS, forces the MS use the FBTS. The MS captured by the attacker who controls what messages go between the MS and BTS as well as messages flowing in the other direction. After capturing MS identity, the attacker will then use this to provide fabricated messages on behalf of a legitimate subscriber.

Attacks on the Anonymity of GSM user

The anonymity in GSM is provided by using temporary identifier TMSI, which is like a nickname of subscriber locally. An attacker may want some subscriber's movements and/or pursue call samples and so must have the IMSI and the TMSI of the MS. This information may also be used to attack other security assets than anonymity, for instance eavesdropping on a specific person. If the attacker can get the IMSI of subscriber or associated current TMSI of a specific person then the anonymity of the user is imperiled.

Passive Monitoring

Every time a MS is powered on, MS is required to introduce itself to the network. This is performed by an IMSI attach. IMSI attach occurs in case of location update. Since the IMSI is not registered in the network and there is not yet any authentication, an encryption cannot be applied. Therefore the IMSI is sent in the clear. An attacker listening to the air traffic can extract the IMSI and associated subscriber's being active.

Passively track GSM users and eavesdropping on the users' permanent identity (IMSI) is possible and easy. This information provides the attacker with a functional IMSI and the knowledge of that the owner of IMSI is in the present area. Passive monitoring is however inefficient and time-consuming since the attacker needs to either wait for MSs to perform IMSI attach when it is powered on or for a database failure to occur in the network, which probably does not happen so frequently.

Active Monitoring

To track a GSM subscriber, the attacker can make use of the identification procedure. The network may initiate an identification procedure, if the network cannot identify the MS using its TMSI. The identification procedure begins with transmitting an IDENTITY REQUEST message to MS so as to ask it to transmit an identification parameter. The network can request IMSI, IMEI or TMSI. Since GSM does not use message authentication to check message origin on the radio link, an attacker with sufficient base station functionality can use these messages to retrieve the same information as a legitimate base station by deceiving the target MSs.

It should be possible to request from a subscriber (whose IMSI is known by attacker) for his/her TMSI abusing the identification procedure. When the attacker knows the IMSI/TMSI bundle, it is possible to locate a specific subscriber. The attacker simply pages the MS with the specific IMSI/TMSI.

Attacks on the Authentication Algorithm

Many GSM operators use the design specification given in the GSM MoU, COMP128, instead of designing their own algorithm for authentication (A3) and session key generation (A8). The difficulty in starting to create new algorithm is that subscribers who bought SIMs before eventual introduction of a different algorithm, are forced to use their old SIMs with the old algorithm. Other reason to change/revise the algorithm is the cost of changing software in database etc. On the other hand, it is possible to utilize new and more secure versions of COMP128 in new SIMs that are given to new subscribers.

The design of COMP128 was never made public, but the design has been reverse engineered and cryptanalyzed. Today, it is quite easy to find software implementation of COMP128 by simple search on internet. Since the GSM specification for SIM cards is widely available, all that is needed to clone a SIM card is the 128 bit COMP128 secret key Ki and the IMSI which is coded in the SIM.

By copying Ki and IMSI into an empty SIM (easy to buy from web) the attacker can authenticate himself to the network as the legitimate subscriber and thus call by charging. The attacker can even, instead of using the subscription, use the captured key Ki for decrypting all the calls from and to the subscriber.

Cloning can be done either by physical access to the SIM to be cloned, or over the air.

Cloning with Physical Access to the SIM Module

If the attacker has physical access to the SIM module, several attacks can be launched in order to clone the module. Some of these attacks base on using flows in the cryptographic algorithm resided in the smart card, while others use vulnerabilities in the smart card itself.

The most popular attack to SIM modules is the attacks to the cryptographic algorithm (COMP128) itself. It is a chosen-challenge attack and use flows in the hashing function to deduce the secret key Ki. The attacker creates a number of specially-chosen challenges and queries the SIM for each one. The SIM applies COMP128 to its secret key and the chosen challenge, returning a response back. After analyzing the responses, the attacker can determine the Ki. The result of this attack is thus that the attacker gains access to the secret key Ki of the MS. The attack exploits a lack of diffusion, which means that some parts of the output hash depend only on some parts of the input to the algorithm. Mounting this attack requires, apart from having physical access to the target SIM, an off-the-shelf smartcard reader, and a computer to direct the operation. The attack requires one to query the SIM about 150,000 times; an average SIM reader can issue 6.25 queries per second, so the whole attack takes approximately 8 hours. By overclocking the SIM or using a higher frequency oscillator on the SIM card reader the processing time could be reduced considerably. This increases however the risk of failure and damage to the original SIM.

Cloning over the Air

The attacker can even perform the attack over the air, making use of a fake base station. Apart from this equipment, the attacker needs to know the target IMSI or TMSI. The captured MSs will be immediately forced to make a location update request which is conducted. After the channel allocation is completed the attacker initiates an authentication process. Immediately after the attacker has a challenge-response pair, he/she initiates a new authentication procedure. The MS is required to respond to every challenge made by the GSM network. This process continues until the attacker has got the required number of pairs to be able to initiate the cloning procedure.

It is assumed that the channel establishment stage only has to be done once. The number of frames exchanged between the network and an MS, for one authentication process, are approximately 66 frames. Since the duration of one TDMA frame is 4.610 ms, the duration of the whole signaling sequence is 4.615 ms/frame x 66 frames = 0.30459 s. The time it takes to get the number of challenge-response pairs needed for the attack can be calculated. It is known that the cryptographic attack requires approximately 150 000 challenge-response pairs. This means that the attack takes approximately 45,689 seconds (150 000 challenges x 0.30459 s), that is approximately 13 hours. This means that the MS has to be available to the attacker over the air for the whole time it takes to gather the information. This is quite unrealistic, because people use their mobiles to make calls or receive calls in addition to the fact that such a bombardment with challenges may cause the battery of the MS to run out, which would make the victim suspicious. To get rid of these problems, the attack can be performed in parts; instead of performing a 13-hour attack, the attacker could interrogate the MS for 30 minutes every day. In that way, the battery would not run out and there would be less risk of making the owner or the legitimate network suspicious.

The defense against cloning over the air is to limit the number of times a SIM can be authenticated to a number significantly smaller than 1,50,000. The SIM locks up if the limit is exceeded. The drawback about this solution is that a new SIM module has to be issued and distributed to the subscriber, which results in costs both for the subscriber and the operator.

Attacks on the Confidentiality of GSM

As mentioned before, the over-the-air privacy of GSM telephone conversations is protected using the A5 stream cipher. This algorithm has two main variants: A5/1 is the "strong" export-limited version used by CEPT-countries, and A5/2 is the "weak" version that has no export limitations. The exact design of both A5/1 and A5/2 was reverse engineered by Briceno from an actual GSM telephone in 1999.

Attacks are classified into three: brute-force attacks, crypto analytical attacks, and non-crypto analytical attacks.

Brute-force Attacks

The confidentiality of GSM is protected by the secrecy of Kc. Kc is 64 bits although the last 10 bits are set to zero. This reduces the key space from 2^{64} to 2^{54}. A5/2 was developed with assistance from the NSA, and can be broken in real time with a work factor of approximately 2^{16}. A5/1, the stronger of the two variants, is however susceptible to attacks that can break it with a work factor of 2^{40}.

Pentium 4 chip has nearly 60 million transistors and the implementation of one set of LFSRs (A5/1) would require about 2000 transistors, 30.000 parallel A5/1 implementations on one chip can be done. If the chip was clocked to 3.2 GHz (a rather ambitious assumption) and each A5/1 implementation would generate one output bit for each clock cycle then it is needed to generate 100+114+114 output bits, hence approximately 10M keys per second per A5/1 implementation can be used. A key space of 2^{54} would thus require about 18 hours, using all of the parallel implementations on the chip. If the attack in the average case succeeds after searching half of the key space, the key is found in about 9 hours. Further optimization by giving up on a specific key after the first invalid key stream bit and distributing the computation between multiple chips will decrease the computation time by several magnitudes. This, still in the worst case, means several hours/many minutes of processing and is far away from a real-time attack. Bear in mind that the complexity of the attack is even greater due to the fact that it is quite difficult to determine when the key is found due to the nature of the plaintext.

To conclude, it is too difficult to succeed in a brute-force attack in real-time, but it is fully possible to find a key given a couple of hours. Entities with enough resources (computation power) can probably cut the processing time greatly.

Even though a brute-force attack may not be used as a real-time attack on the A5 algorithm, it could easily be used to find the key used in a specific conversation "offline". The attacker intercepts and records the interesting conversation and decrypts it at a later time.

Crypto Analytical Attacks

There exist several crypto analytical attacks against the algorithms protecting different aspects of GSM. The algorithm used by many operators to authenticate subscribers (COMP128) is broken due to flaws in the design of the hash function. The result is the ability of intruders to clone subscriptions either by having physical access to the target SIM or over the air. The most popular attack requires physical access to the SIM to clone and is completed in about 8 hours. It can be speeded up with the risk of damaging the SIM. The most efficient way to clone a GSM smartcard is a partitioning attack proposed by a team from IBM. It requires challenging the target SIM only 8 times in the best case, which means that cloning can be done in minutes or even seconds. The equipment needed to mount this attack (a specially designed smartcard reader and software) is however only available in laboratories yet. Newer versions of COMP128 has been developed and distributed. It is however not known to what extent these stronger versions have been adapted by operators. A guess is that many operators still use the old algorithm due to the costs involved in upgrading. What is known for sure is that users who had COMP128 inside their SIMs when they bought a subscription are still using COMP128.

There exist several crypto analytical propositions on how to attack the encryption algorithms used for confidentiality protection that break these algorithms in real-time. Several attacks against A5/1 and A5/2 exist, although most of them have only theoretical value. Most of the attacks require that the attacker knows portions of the key stream. It is possible to obtain small portions of plaintext because the attacker often knows the structure and content of the signaling messages (especially if the attacker is impersonating the network to the victim MS and is thereby able to query the MS for information) in addition to the fact that channel coding is applied to the data before encryption. An attacker mounting a man-in-the-middle attack may ask the victim subscriber to transmit certain signaling messages (of which the content is known or almost known) after encryption has started. The attacker then has access to the cipher-text in addition to the known portions of the plaintext and can thereby derive portions of the key stream used in the encryption process. It is, however, hard to obtain the amounts of the known plaintext that some of these attacks require. The attack that requires least known plaintext is an attack against A5/2. It requires that the attacker knows the plaintext of two frames approximately six seconds apart from each other and finds the session key in about 10 ms. The known plaintext requirement may be possible to satisfy using the method mentioned earlier, therefore this attack on A5/2 has been used in one of the attacks on confidentiality. It is worth mentioning that it only took a couple of hours to crack A5/2, which illustrates the weaknesses of this algorithm.

The most recent attack on A5/1 is a cipher-text-only attack. This is an impressive attack only requiring knowledge of a small number of encrypted frames, enabling the attacker to listen to the "encrypted" conversation data, in real-time. Further the authors of propose a ciphertext-only attack on A5/2 that improves the previous attack on A5/2 to a ciphertext-only attack. The problem of known plaintext is no longer a concern using this attack. This is however an attack requiring huge amounts of computation power. Figure below gives an idea of the computation requirements in the proposed ciphertext-only attack on A5/1.

Three Possible Tradeoff Points in the Attacks on A5/1

Available data (ciphertext)	Pre-processing steps	Number of PCs to complete pre-processing in one year	Number of 200 GB disks	T	Number of PCs to complete attack in real-time
2^{12}(appr 5 min)	2^{52}	140	22	2^{28}	1
$2^{6.7}$(appr 8 sec)	2^{41}	5000	176	$2^{32.6}$	1000
$2^{6.7}$(appr 8 sec)	2^{42}	5000	350	$2^{30.6}$	200
2^{14}(appr 20 min)	2^{35}	35	3	2^{30}	1

Since this is a ciphertext only attack, no plaintext is required in order to find the session key in real-time. However, the computation and storage requirements for this attack are very high making it very unlikely that an individual hacker would have the needed resources to mount the attack. The requirements for the ciphertext-only cryptanalysis of A5/2 are however fulfilled by most personal computers of today.

Non-crypto Analytical Attacks

It is common knowledge that most GSM mobile phones can communicate with most different base stations and networks. This is possible because all of the different manufacturers follow the specifications and standards of how GSM should function. These specifications are developed by the European Telecommunications Standards Institute (ETSI). It is possible to study the specifications on how the communication between the network and the MS is conducted, and get detailed information on the communication protocols and mechanisms used when a MS is to be authenticated by the network.

The same key Kc is used for the different encryption algorithms A5/1, A5/2, and A5/3. This means that breaking one of this three algorithms and retrieving the session key threatens the confidentiality of the conversation even when the stronger versions of the algorithm are used later.

A base station does not need to authenticate itself to the MS it is communicating with. Furthermore messages are not authenticated and their integrity is not protected.

Denial of Service (DoS) Attacks

DoS attacks can be performed by physically disturbing radio signals or by logical means.

Denial of Service – Physical Intervention

The physical attacks are the most straight forward attacks. The attacker prevents user or signaling traffic from being transmitted on any system interface, whether wired or wireless, by physical means. An example of physical intervention on a wired interface is wire cutting. The attacker could for example cut the wire leaving a base station. An example of physical intervention on a wireless interface is jamming. Having the equipment that jam GSM radio signals is sufficient. The equipment is placed in the area where traffic is to be disturbed and the GSM equipment within the device's range will not function properly. Note that the frequency hopping makes the jamming more difficult than usual.

There are examples of jamming causing problems for GSM operators. Recently a GSM operator in Moldova suffered heavily from jamming activities that effectively caused a drop rate of lost calls of about 7 %. The operator and the authorities had major problems in stopping the attacks.

Denial of Service – Logical Intervention

An attacker can perform DoS attacks by logical means also as the following examples show:

- The attacker spoofs a de-registration request (IMSI-detach) to the network. The network de-registers the subscriber from the visited location area and instructs the HLR to do the same. The user is subsequently unreachable for other subscribers. The attacker needs a modified MS and the IMSI of the user to de-register.

- The attacker spoofs a location update request in a different location area from the one in which the subscriber is roaming. The network registers the subscriber in the new location area and the target user will be paged in that new area. The user is subsequently unreachable for mobile terminated services.

- An attacker in possession of a modified base station, transmitting the base channel with higher signal strength will force the MSs in the area to camp on the radio channels of the false base station, making them unreachable for the serving network.

Some useful solutions against Attacks

Regardless of security improvements in generation networks, it is necessary to provide solutions to improve the security of the currently available 2G systems. Some practical solutions are discussed in the below.

Using Secure Algorithms for A3/A8 Implementations

This can thwart the dangerous SIM card cloning attack. This solution is profitable since the network operators can perform such improvement themselves and without any need to the software and hardware manufacturers or the GSM consortium. However, this solution requires providing and distributing new SIM cards and modifying the software of the HLR. Currently, both COMP128-2 and COMP128-3 algorithms thwart the SIM card cloning and over-the-air cracking of Ki. Since COMP128-3 enhances the effective key length of the session key to further 10 bits, it allows the deployed cryptographic algorithm to have its nominal security. Although it is soon to judge on the real security of COMP128-2 and COMP128-3, they have apparent advantages over the traditional COMP128-1 that its SIM cloning apparatus are available at very low prices.

Using Secure Ciphering Algorithms

Operators can use newer and more secure algorithms such as A5/3 provided that such improvements are allowed by the GSM consortium. The deployed cryptographic algorithms should be implemented on both BTS and mobile phones. Any change to the cryptographic algorithms requires agreement and cooperation of software and hardware manufacturers since they should perform the appropriate changes to their products. Since the cryptographic algorithms should be implemented on the cellular phones, the agreement of mobile phone manufacturers is also required. However, a lonely upgrading of the deployed cryptographic algorithms cannot be so useful. Even though the ciphering algorithms are replaced with the strongest ones, the attacker can simply impersonate the real network and force MS to deactivate the ciphering mode so it is also necessary to modify the authentication protocols.

End-to-end Security

The best, easiest, and most profitable solution is to deploy the end-to-end security or security at the application layer. Most of GSM security vulnerabilities (except SIM cloning and DoS attacks) do not aim ordinary people, and their targets are usually restricted to special groups so it is reasonable and economical that such groups make their communications secure by the end-to-end security. Since the encryption and security establishment is performed at the endentities, any change to the GSM hardware will not be required. In this way, even if the conversation is eavesdropped by the police or legal organizations, they cannot decrypt the transmitted data without having the true ciphering key, provided that a secure enough cryptographic algorithm is deployed.

Therefore, in order to avoid illegal activities, it should be transparent to both GSM operator and service provider.

CDMA Security

As a digital communication system, security management is very straightforward to be achieved for CDMA. The security architecture for CDMA system defines security features intended to meet certain threats. It also sets up required security services for example network access security, to provide confidentiality of user identity and that of the user and signaling data. The integrity protection of important signaling data, authentication of user and network, and identification of Mobile Station (MS) is also done by this feature. The feature network domain security enables different nodes in the provider domain to securely exchange signaling data and also offers protection against attacks on the wire line network. The user domain security guarantees that only authorized access to Universal Subscriber Identity Module (USIM) is made. On the other hand, application domain security is used to enable applications in the user and provider domains to securely exchange messages.

The figure below illustrates the security parameters and algorithms which are used in the fundamental CDMA security system:

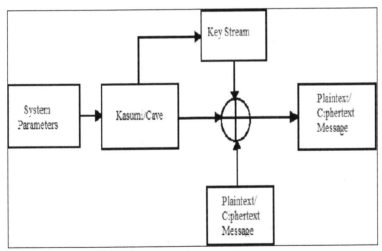

CDMA security procedures.

The authentication algorithm in the fundamental CDMA system is enclosed within the smart card. The individual key for each IMSI must be chosen to be random, and must be safeguarded in order to prevent the user from being duplicated. Throughout the security process Ki should be protected.

The Basic Cdma-Aka Protocol

This protocol was designed within USECA and one of its design criteria, the compatibility with the security architecture of second generation systems. Furthermore, different methods for authentication and key agreement (AKA) are described. A vital characteristic of an AKA protocol is the goal the protocol achieves.

Entity authentication: for entity authentication of users to the network operator, a protocol employing random challenges as time variant parameters provides a guarantee to the network operator that the evidence was created during the current protocol run. A sequence number as time variant parameter only provides a guarantee to the network operator that the evidence was not employed in a prior protocol run.

Assurance of key freshness: to the network operator alludes to the fact that the network operator (HLR/AuC) can be sure that the keys obtained in the course of the AKA protocol were not employed before the current protocol run.

The following figure depicts the AKA protocol in the basic CDMA cellular standard:

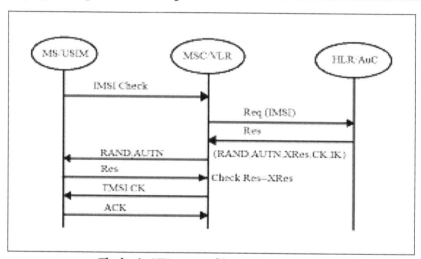

The basic AKA protocol in CDMA standard.

Key confirmation: for a user to the HLR/AuC provides an assurance to the HLR/AuC that the specific user holds the correct parameter(s) to derive the agreed keys. Key confirmation is the stronger aim and provides assurance to HLR/AuC that the specific user holds the derived key itself. Goals of CDMA-AKA:

- Entity authentication of MS/USIM to HLR/AuC.

- Implicit key authentication of HLR/AuC to MS/USIM.

- Implicit key authentication of MS/USIM to the HLR/AuC.

- Assurance of key freshness to MS/USIM and HLR/AuC.

- Key confirmation from MS/USIM to HLR/AuC and vice-versa.

- Confidentiality of the user identity and other user related information from which the user identity may be obtained on the interface between MS/USIM and HLR/AuC. Also, if the transmission of the identity is not a component of the AKA and is instead provided by alternate means, the AKA must not prevent the provision of confidentiality of the user identity on the air interface. If the CDMA-AKA protocol does not offer key confirmation, then the usage of the agreed keys after a successful operation of the AKA, will (in case of a failure in one of the agreed keys) lead to problems in the messages protected by these keys and therefore to a break in the transmission of user/signaling data which otherwise could have been detected beforehand.

Limitations of the Basic Cdma-Aka Protocol

Some of the most serious attacks discussed were based on the accessibility of supposed false base stations. A part of this problem is the so-called IMSI catching, which is a danger to the discretion of the user identity over the air interface. An additional type of attack analyzed was linked to an attacker taking control of the user's services. One vital result of the threat analysis was that some signaling elements were regarded as being sensitive and therefore having to be integrity protected. One of these signaling elements is the secure mode command, which establishes whether or not ciphering is enabled and the ciphering and integrity algorithm to be used. Another example is the set of MS capabilities transmitted from the MS/USIM to the network operator, including authentication and key agreement mechanisms, ciphering algorithm and message authentication function capabilities. Corresponding contributions on threats and countermeasures were forwarded to 3GPP and formed a major input for the 3GPP technical report.

CDMA security system does not define a standard authentication algorithm; rather, it allows operators to pick their own versions, which conform to the published standards. However, in order to aid operators, guidelines are available as to how to develop an appropriate algorithm. Threats of medium significance include eavesdropping signaling or control data on the wireless or other interfaces; camouflaging as another user; manipulation of the terminal or USIM behavior by camouflaging as the creator of applications and/or data; camouflaging as a serving network; integrity of data on a terminal or USIM. As can be deduced from the above enumeration, the results of the threat analysis categorizes the main threats as coming from camouflaging as other users to get illegal access to services, eavesdropping which may result in the compromise of user data traffic privacy, or of call-related information like dialed numbers, location data, etc., and subscription fraud where subscribers take advantage of the services with heavy usage without any plan to pay.

What is new is the acknowledgment of threats which exploit more sophisticated, active attacks to achieve the eavesdropping or camouflaging.

These comprise of attacks which involve the manipulation of signaling traffic on the radio interface and where the intruder camouflages as a base station. Furthermore, attention is now not only focused on radio interface attacks, but also on other parts of the system.

3G Security

Security Mechanisms in 3G – UMTS

Since the underlying technology is the same, security features of one architecture are applicable to others as well. 3G - UMTS, the most popular of the architectures builds upon the security features of 2G systems so that some of the robust features of 2G systems are retained. The aim of the 3G security architecture is to improve on the security of 2G systems. Any holes present in the 2G systems are to be addressed and fixed. Also, since many new services have been added to 3G systems, the security architecture needs to provide security for these services.

3G Security Architecture

There are five different sets of features that are part of the architecture:

- Network Access Security: This feature enables users to securely access services provided by the 3G network. This feature is responsible for providing identity confidentiality, authentication of users, confidentiality, integrity and mobile equipment authentication. User Identity confidentiality is obtained by using a temporary identity called the International Mobile User Identity. Authentication is achieved using a challenge response method using a secret key. Confidentiality is obtained by means of a secret Cipher Key (CK) which is exchanged as part of the Authentication and Key Agreement Process (AKA). Integrity is provided using an integrity algorithm and an integrity key (IK). Equipment identification is achieved using the International Mobile Equipment Identifier (IMEI).

- Network Domain Security: This feature enables nodes in the provider domain to securely exchange signaling data, and prevent attacks on the wired network.

- User Domain Security: This feature enables a user to securely connect to mobile stations.

- Application Security: This feature enables applications in the user domain and the provider domain to securely exchange messages.

- Visibility And Configurability Of Security: This feature allows users to enquire what security features are available.

The UMTS Authentication and Key Agreement (UMTS AKA) mechanism is responsible for providing authentication and key agreement using the challenge/response mechanism. Challenge/Response is a mechanism where one entity in the network proves to another entity that it knows the password without revealing it. There are several instances when this protocol is invoked. When the user first registers with the network, when the network receives a service request, when a location update is sent, on an attach/detach request and on connection reestablishment. The current recommendation by 3GPP for AKA algorithms is MILENAGE. MILENAGE is based on the popular shared secret key algorithm called AES or Rijndael. Readers interested in the AES algorithm are encouraged to look at . AKA provides mutual authentication for the user and the network. Also, the user and the network agree upon a cipher key (CK) and an integrity key (IK) which are used until their time expires.

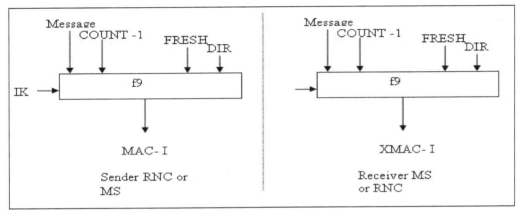

Signaling Data Integrity Mechanism.

Control Signaling Communication between the mobile station and the network is sensitive and therefore its integrity must be protected. This is done using the UMTS Integrity Algorithm (UIA) which is implemented both in the mobile station and the RNC. This is known as the f9 algorithm. Figure shows how this algorithm is applied. First, the f9 algorithm in the user equipment calculates a 32 bit MAC-I for data integrity using the signaling message as an input parameter. This, along with the original signal message is sent to the RNC, where the XMAC-I is calculated and then compared to the MAC-I. If both are same, then we know that the integrity of the message has not been compromised.

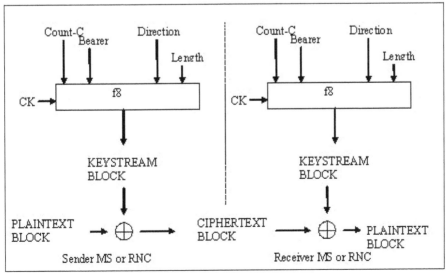

Air Interface confidentiality.

The confidentiality algorithm is known as f8 and it operates on the signaling data as well as the user data. Figure shows how this algorithm is applied. The user's device uses a Cipher Key CK and some other information and calculates an output bit stream. Then this output stream is xored bit by bit with the data stream to generate a cipher stream. This stream is then transmitted to the RNC, where the RNC uses the same CK and input as the user's device and the f8 algorithm to calculate the output stream. This is then xored with the cipher stream to get the original data stream.

For more information on the inputs to the f8 and f9 algorithms, please refer to. A block cipher known as the KASUMI cipher is central to both the f9 and the f8 algorithm. This cipher is based on the feistel structure using 64 bit data blocks and a 128 bit key.

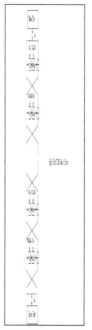

KASUMI Block cipher.

It has eight rounds of processing, with the plain text (can be any form of data) as input to the first round and the cipher text the result after the last round. An encryption key is used to generate round keys (KLi,KOi,KIi) for each round i. Each round calculates a separate function since the round keys are different. The same algorithm is used for encryption and decryption. The KASUMI cipher is based on the MISTY1 cipher which was chosen by 3GPP due to its proven security against many advanced cipher breaking techniques. It has been optimized for hardware implementation which is important concerning the hardware constraints of cellular devices, such as limited power and limited memory. As shown in the Figure, the function f consists of subfunctions FLi and FOi. FL is a simple function consisting of shifts and logical operations. The FO function is much more complicated and is itself based on the fiestel structure and consists of three rounds. Anyone interested in the details of the KASUMI algorithm are encouraged to look at.

Wireless Application Protocol (WAP)

Since one of the most important services provided by 3G systems is access to the Internet, it is important to understand the security mechanisms of the protocol used to access the Internet. WAP is an open specification which enables mobile users to access the Internet. This protocol is independent of the underlying network e.g. WCDMA, CMDA 2000 etc. and also independent of the underlying operating system e.g. Windows CE, PALM OS etc. The first generation is known as WAP1 which was released in 1998. WAP1 assumes that the mobile devices are low on power and other resources. And therefore the devices can be simple while sharing the security responsibilities with the gateway devices. The second generation is known as WAP2 and was released in 2002. WAP2 assumes that the mobile devices are powerful and can therefore directly communicate with the servers. Figures show the protocol stack for WAP1 and WAP2 respectively.

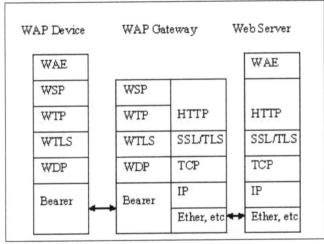

WAP1 Protocol Stack.

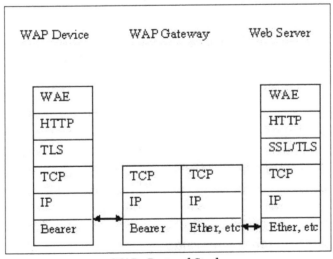

WAP2 Protocol Stack.

A brief description of each layer is as follows:

- Wireless Application Environment (WAE): This provides an environment for running web applications or other WAP applications.

- Wireless Session Protocol (WSP): This is similar to the HTTP protocol and provides data transmissions with small sizes so that WAP1 clients can process the data with less complexity.

- Wireless Transaction Protocol (WTP): This is responsible for providing reliability.

- Wireless Transport Layer Security (WTLS): This is responsible for providing security features such as authentication, confidentiality, integrity etc. between a WAP1 client and the WAP gateway.

- Wireless Datagram Protocol (WDP): This provides the underlying transport service.

- Hypertext Transfer Protocol (HTTP): A standard protocol used to transmit web pages.

- Transport Layer Security (TLS): This layer provides security features such as authentication, confidentiality, integrity etc. In WAP1, this is between the WAP1 gateway and the server. In WAP2 this is between the WAP2 client and the server.

- Transport Control Protocol (TCP): Standard transport protocol used to provide reliability over IP.

- Internet Protocol (IP): Protocol used to route data in a network.

- Bearer Protocol: This is the lowest level protocol and can be any wireless technique such as GSM, CDMA etc.

- Cipher Suite in WTLS: This suite provides a key-establishment protocol, a bulk encryption algorithm and a MAC algorithm. In SSL/TLS these are used together, in WTLS each can be used independently.

- Key Exchange Suite: This protocol is responsible for establishing a secret key between a client and the server. An example of is the RSA key suite, which consists of the following steps: the WAP gateway sends a certificate consisting of the gateway's RSA public key and signed by the certification authority's private key. The client checks the validity of the certificate authority's signature. If invalid, the communication is aborted. If valid, the user generates a secret value, encrypts it with the gateway's public key. Both sides can then calculate their common keys using the secret value.

- Bulk Encryption And MAC Suite: Bulk encryption is used for data confidentiality and the MAC is used for integrity. The common key that we calculated in the key exchange suite can be used for both purposes. For bulk encryption, algorithms such as DES, 3DES, IDEA and RC5 are used. For integrity WTLS uses the HMAC algorithm which uses either SHA-1 or MD5 twice.

- WAP-Profiled TLS: WAP2 uses the WAP profiled TLS which consists of a cipher Suite, authentication suite, tunneling capability and session identification and session resume. Cipher suite consists of key establishment (e.g. RSA), encryption (e.g. DES) and integrity (SHA-1 for MAC calculation). A session identifier is chosen by the server to identify a particular session with the client. Server and Client authentication is done using certificates similar to WTLS. Tunneling is a mechanism set up between the client and the server, so that they can communicate even if the underlying network layers are different.

- WAP Identity module: WIM (WAP Identity Module) is a method of identification in WAP. This enables the device to separate its identification from WAP. So a device can be updated without any changes made to the telephone number or billing information. WIM provides operations such as key generation, random numbers, signing, decryption, key exchange, storing certificates etc.

4G/LTE Network Security

Mobile network operators (MNOs) must grapple with complex security management in fourth generation Long Term Evolution (4G LTE) deployments. The security architecture of 4G LTE may lull MNOs into a sense of complacence that the technology intrinsically addresses security in LTE operations. 4G LTE has known security vulnerabilities. Besides inherent LTE vulnerabilities, 4G LTE includes long standing internet protocol (IP) based security weaknesses. The third generation partnership project (3GPP) has included security in their system architecture evolution (SAE) from inception, yet there are numerous security considerations deferred to the MNO. In terms of service delivery and operations MNOs are left to manage both LTE and IP based security vulnerabilities. This leads to complex security management requirements for MNOs. LTE is designed with strong cryptographic techniques, mutual authentication between LTE network elements with security mechanisms built into its architecture. However, trusted industry organisations have identified security vulnerabilities that should be assessed by virtue of network deployment. With the emergence of the open, all IP based, distributed architecture of LTE, attackers can target mobile devices and networks with spam, eavesdropping, malware, IP-spoofing, data and service theft, DDoS attacks and numerous other variants of cyber-attacks and crimes. MNOs are focused on increasing business profitability by 4G deployments, and are the first point of contact, for subscribers in the event of security or privacy breaches. To protect profit dollars from being spent on recovery and remediation from security breaches, MNOs should keep abreast of prevalent security risks in both LTE and IP, the evolving security threatscape and actively invest in preventative security measures.

4G/LTE Security Controls

Abstraction layers are inserted in the 4G/LTE architecture in terms of the unique identifiers (IDs) for smartphones (i.e., UEs). A temporary unique ID is used on the SIM card to prevent attackers from stealing identifiers. Another technique for improving 4G security is adding protected singling between the UE and MME. Security mechanisms are utilized to secure the connections between 4G networks and secure non-4G networks using key management authentication protocols. Although several security controls are used for 4G/LTE wireless technology, its design, which is based on an open-IP architecture, and the sophistication of APT hackers make security and privacy of 4G/LTE systems challenging.

4G/LTE Security Requirements

In order to secure mobile devices that use 4G/LTE wireless technologies, there should be protection for the connections between the UEs and MMEs and between elements in the wireline networks and mobile stations. For satisfying these requirements, the 4G/LTE security is significantly improved by adding (1) advanced key hierarchy, (2) protracted authentication and key agreement, and (3) additional interworking security for the NEs. The requirements are classified into key building blocks and LTE end-to-end security.

- Key building blocks include the following elements:

  - Key security and hierarchy: LTE has five key strategies used for connections of the EPS and E-UTRAN. The keys are declared as follows: (1) KANS encryption and integrity keys are used to protect non-access stratum (NAS) traffic between the UE and MME, (2) a KUP encryption is used to encrypt traffic between the UE and eNodeB, and (3) KPRC encryption and integrity keys are used to secure the Radio Resource Control (RRC) between the UE and eNodeB.

  - Key management: LTE key management comprises three functions: key establishment, distribution, and generation. It is essential that 4G/LTE wireless technology has key management mechanisms that prevent stealing keys, as mobile devices with IP-based infrastructure can frequently access different wireless networks. An Authentication and Key Agreement (AKA) process is utilized for establishing and verifying keys in 4G/LTE systems.

  - Authentication, encryption, and integrity protection: LTE depends on using regular updating of the authentication process by exchanging sequence numbers in the messages of encryption mechanisms. The IPsec protocol and tunnels are also used for asserting the confidentiality of users' data while transmitting traffic between LTE nodes.

- ○ Unique user identifiers: LTE has several user identifier mechanisms that thwart attackers from learning mobile user identities; therefore, attackers cannot track user profiles or launch denial of service (DoS) attacks against users. The identifier mechanisms contain the following: (1) international mobile equipment identifier (IMEI) which is a permanent unique identifier for each mobile, (2) M-TMSI which is a temporary identifier that defines the UE inside the MME, and (3) cell radio network temporary identifier (C-RN-TI) which is a unique and temporary UE identity when a UE is connected with a cell.

- LTE end-to-end security involves the following elements:

  - ○ Authentication and Key Agreement (AKA): The foundation of LTE security is authenticating the UEs and wireless networks. This can be accomplished using the AKA process which asserts that the serving network authenticates the identity of a user and the UE certifies the network signature. The AKA creates encryption and integrity keys applied for originating various session keys for ensuring the 4G/LTE security and privacy.

  - ○ Confidentiality and integrity of signaling: Security of network access control planes is achieved when the RCC and NAS layer signaling is encrypted and integrity protected. Ciphering and integrity protection of LTE RRC signaling is executed at the packet data convergence protocol (PDCP) layer, whereas the NAS layer attains the protection by encrypting the NAS-level signaling. This protection cannot be uniquely performed for each UE connection, but it runs across trusted connections between AGW and eNodeB.

  - ○ User plane confidentiality: LTE has a security feature for user plane via encrypting data/voice between the UE and eNodeB. Encryption is executed at the IP layer by utilizing IPsec-based tunnels between AGW and eNodeB, but no integrity protection is offered for the user plane due to performance and efficiency considerations. The PDCP layer is used for enabling encrypting/decrypting the user plane while transmitting traffic between the eNodeB and UE.

Cyberattacks and Countermeasure Techniques

4G/LTE wireless technology faces different types of cyberattacks that could affect integrity, privacy, availability, and authentication, as described below:

- Privacy attacks: Attacks against the privacy of mobile users' data attempt to expose sensitive data/multimedia of users. A man-in-themiddle (MITM) attack is the most serious privacy attacks in wireless networks that depend on a false base station attack when anomalous third-party masquerades its base transceiver station. Privacy-preserving authentication and encryption mechanisms have been widely used to protect wireless networks against the MITM attacks.

- Integrity attacks: Attacks against integrity attempt to modify exchanging data between the 4G access points and mobile users. Cloning attacks based on the MITM and message modification scenarios are the major integrity attacks that alter mobile user information. Authentication and privacypreserving mechanisms with hash functions have been broadly used for securing 4G wireless networks against integrity attacks.

- Authentication attacks: Attacks against authentication attempt to disturb the client-to-server and/or server-toclient authentication process. The password reuse, brute force, password stealing, and dictionary attacks are popular wireless hacking schemes that interrupt the password-based authentication. In the hacking schemes, an attacker can pretend to be a legal user and try to log in to a server by guessing various words as a password from a dictionary. Encryption and authentication techniques have been utilized for preventing such kind of attacks from 4G wireless networks.

- Availability attacks: Attacks against availability try to make services unavailable, such as the service of data routing. The first in first out (FIFO) and DoS attacks can be launched by flooding massive malicious actions to 4G wireless victims for disrupting their computational resources. Firewall and intrusion detection systems have been usually used for defending against these attacks.

4G/LTE Challenges and Future Trends

Despite a plethora of research and technical studies that have been conducted for securing 4G/LTE wireless networks, there are several challenges that should be the focus of researchers in future.

- Designing a flexible and scalable 4G/LTE architecture that can address security and privacy issues is an arduous task. There are multiple devices and systems that are usually connected with 4G networks that result in vulnerabilities and loopholes in networks.

- Discovering DoS attacks that attempt to violate 4G wireless networks, as hackers frequently establish new sophisticated variants against eNodeB, UE, and discontinuous reception services.

- Location tracking denotes tracing the UE presence in a specific cell(s). While many portable devices could link to a 4G/LTE wireless network, ensuring that location tracks of the devices are not breached is still a challenging issue, due to the considerations of operability and scalability.

- The utilization of an effective 4G wireless Software Dened Network (SDN) is a challenge. More specifically, there are technical gaps in the network scalability, security, and privacy issues with the SDN.

- Trusted connections through 4G networks in the existence of eavesdroppers are the issues. Especially, when 4G wireless technology is used in the Internet of Things, it requires new cryptographic mechanisms that provide protection and integrity for smartphones and computer systems.

- Instead of individual security techniques, a systematic security and privacy protection strategies are required for 4G/LTE wireless connections while connecting with cloud and edge computing paradigms. This will provide valid security mechanisms, for example, trust models, device security, and data assurance techniques.

Challenges	Cyber-attacks	Security and privacy methods
A resilient 4G/LTE architecture.	Privacy attacks: replay, MITM, impersonation, collaborated, tracing, spoong, privacy violation, masquerade.	Privacy-preservation, authentication and encryption mechanisms.
Tracking locations of devices.	Integrity attacks: cloning, spam, message blocking, message modification attack, message, insertion, tampering.	Hashing and encryption, and authentication and privacy-preserving methods.
An effective 4G/LTE wireless Software Defined Network (SDN).	Availability attacks: FIFO, redirection, physical attack, skimming, and free-riding.	Firewall systems, signature-based and anomaly-based systems.
Collaborative 4G/LTE security and privacy approaches operate on cloud and edge paradigms.	Authentication attacks: password reuse, password stealing, dictionary, brute force, desynchronization, forgery attack, collision, stolen smart card.	Encryption and authentication techniques.

5G Security

Attacks and Security Services in 5G Wireless Networks

Due to the broadcast nature of the wireless medium, wireless information transmission is vulnerable to various malicious threats. Four types of attacks, i.e., eavesdropping and traffic analysis, jamming, DoS and DDoS, and MITM, in 5G wireless networks. We also introduce four security services including authentication, confidentiality, availability, and integrity.

Attacks in 5G Wireless Networks

Figure illustrates all four attacks, each of which is individually discussed in the following three aspects, type of the attack (passive or active), security services provided to fight against this attack, and the corresponding methods applied to avoid or prevent this attack. We focus on security attacks at the PHY layer and MAC layer, where the key difference on security between wireless and wire-line networks occur.

- Eavesdropping and Traffic Analysis: Eavesdropping is an attack that is used by an unintended receiver to intercept a message from others. Eavesdropping is a passive attack as the normal communication is not affected by eavesdropping, as shown in figure. Due to the passive nature, eavesdropping is hard to detect. Encryption of the signals over the radio link is most commonly applied to fight against the eavesdropping attack. The eavesdropper can not intercept the received signal directly due to the encryption. Traffic analysis is another passive attack that an unintended receiver uses to intercept information such as location and identity of the communication parties by analyzing the traffic of the received signal without understanding the content of the signal itself. In other word, even the signal is encrypted, traffic analysis can still be used to reveal the patterns of the communication parties. Traffic analysis attack does not impact the legitimate communications either.

  Encryption method used to prevent eavesdropping is heavily dependent on the strength of the encryption algorithm and also on the computing capability of the eavesdropper. Due to the quick escalation of computing power and booming of advanced data analysis technologies, eavesdropper can take the advantage of the new technologies in theirs attacks. The existing mechanisms to tackle eavesdropping face a big challenge as many of them assume a small number of simultaneous eavesdroppers with low computing capability and low data analysis capability. Moreover, some technologies applied to 5G wireless networks such as HetNet may further increase the difficulty to fight against eavesdroppers. In general the new characteristics of 5G wireless networks lead to many more complicated scenarios to cope with eavesdroppers, for example, in eavesdroppers with multiple antennas are considered. As cryptographic methods to tackle eavesdropping have been extensively investigated in the past and are considered rather mature, most recently, PLS research to tackle eavesdropping has been paid more and more attentions.

- Jamming: Unlike eavesdropping and traffic analysis, jamming can completely disrupt the communications between legitimate users. Figure is an example for jamming attack. The malicious node can generate intentional interference that can disrupt the data communications between legitimate users. Jamming can also prevent authorized users from accessing radio resources. The solutions for active attacks are normally detection based.

  Spread spectrum techniques such as direct sequence spread spectrum (DSSS) and frequency hopping spread spectrum (FHSS) are widely used as a secure communication method to fight against jamming at the PHY layer by spreading the signals over a wider spectral bandwidth. However, DSSS and FHSS based anti-jamming schemes may not fit into some applications in 5G wireless networks. In a pseudorandom time hopping anti-jamming scheme is proposed for cognitive users to improve the performance compared to FHSS. Due to the characteristics of jamming, detection is possible. In a resource allocation strategy is proposed between a fusion center and a jammer. Resource allocation is applied to improve the detection to achieve a better error rate performance.

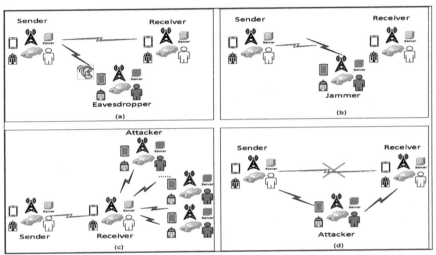

Attacks in 5G wireless networks (a). Eavesdropping; (b). Jamming; (c). DDoS; (d). MITM.

- DoS and DDoS: DoS attacks can exhaust the network resources by an adversary. DoS is a security attack violation of the availability of the networks. Jamming can be used to launch a DoS attack. DDoS can be formed when more than one distributed adversary exists. Figure shows a DDoS model. DoS and DDoS are both active attacks that can be applied at different layers. Currently, detection is mostly used to recognize DoS and DDoS attacks. With a high penetration of massive devices in 5G wireless networks, DoS and DDoS will likely become a serious threat for operators. DoS and DDoS attacks in 5G wireless networks can attack the access network via a very large number of connected devices. Based on the attacking target, a DoS attack can be identified either as a network infrastructure DoS attack or a device/user DoS attack. A DoS attack against the network infrastructure can strike the signaling plane, user plane, management plane, support systems, radio resources, logical and physical resources. A DoS attack against device/user can target on battery, memory, disk, CPU, radio, actuator and sensors.

- MITM: In MITM attack, the attacker secretly takes control of the communication channel between two legitimate parties. The MITM attacker can intercept, modify, and replace the communication messages between the two legitimate parties. Figure shows a MITM attack model. MITM is an active attack that can be launched in different layers. In particular, MITM attacks aim to compromise data confidentiality, integrity, and availability. Based on the Verizon's data investigation report, MITM attack is one of the most common security attacks. In the legacy cellular network, false base station based MITM is an attack that the attacker forces a legitimate user to create a connection with a fake base transceiver station. Mutual authentication between the mobile device and the base station is normally used to prevent the false base station based MITM.

Security Services in 5G Wireless Networks

The new architecture, new technologies, and use cases in 5G wireless networks bring in new features and requirements of security services.

- Authentication: There are two kinds of authentications, namely, entity authentication and message authentication. Both entity authentication and message authentication are important in 5G wireless networks to tackle the previous mentioned attacks. Entity authentication is used to ensure the communicating entity is the one that it claims to be. In the legacy cellular networks, mutual authentication between user equipment (UE) and mobility management entity (MME) is implemented before the two parties communicating to each other. The mutual authentications between UE and MME is the most important security feature in the traditional cellular security framework. The authentication and key agreement (AKA) in 4G LTE cellular networks is symmetric-key based. However, 5G requires authentication not only between UE and MME but also between other third parties such as service providers Since the trust model differs from that used in the traditional cellular networks, hybrid and flexible authentication management is needed in 5G. The hybrid and flexible authentication of UE can be implemented in three different ways: authentication by network only, authentication by service provider only, and authentication by both network and service provider. Due to the very high speed data rate and extremely low latency requirement in 5G wireless networks, authentication in 5G is expected to be much faster than ever. Moreover, the multi-tier architecture of the 5G may encounter very frequent handovers and authentications between different tiers in 5G. In to overcome the difficulties of key management in HetNets and to reduce the unnecessary latency caused by frequent handovers and authentications between different tiers, a SDN enabled fast authentication scheme using weighted secure-context-information transfer is proposed to improve the efficiency of authentication during handovers and to meet 5G latency requirement. To provide more security services in 5G wireless networks, in, a public-key based AKA is proposed.

With the various new applications in 5G wireless networks, message authentication becomes increasingly important. Moreover, with the more strict requirements on latency, spectrum efficiency (SE), and EE in 5G, message authentication is facing new challenges. In an efficient Cyclic Redundancy Check (CRC) based message authentication for 5G is proposed to enable the detection of both random and malicious error without increasing bandwidth.

- Confidentiality: Confidentiality consists of two aspects, i.e., data confidentiality and privacy. Data confidentiality protects data transmission from passive attacks by limiting the data access to intended users only and preventing the access from or disclosure to unauthorized users. Privacy prevents controlling and influencing the information related to legitimate users, for example, privacy protects traffic flows from any analysis of an attacker. The traffic patterns can be used to diagnose sensitive information, such as senders/receivers location, etc. With various applications in 5G, there exist massive data related to user privacy, e.g., vehicle routing data, health monitoring data, and so on.

Data encryption has been widely used to secure the data confidentiality by preventing unauthorized users from extracting any useful information from the broadcast information. Symmetric key encryption technique can be applied to encrypt and decrypt data with one private key shared between the sender and the receiver. To share a key between the sender and the receiver, a secure key distribution method is required. Conventional cryptography method is designed based on the assumption that attackers have limited computing capabilities. Thus it is hard to fight against attackers who are equipped with powerful computing capabilities. Rather than relying solely upon generic higher-layer cryptographic mechanisms, PLS can support confidentiality service against jamming and eavesdropping attacks. Besides the data services of 5G, users start to realize the importance of privacy protection service. Privacy service in 5G deserves much more attention than in the legacy cellular networks due to the massive data connections. Anonymity service is a basic security requirement in many user cases. In many cases, privacy leakage can cause serious consequences. For examples, health monitoring data reveals the sensitive personal health information; vehicle routing data can expose the location privacy. 5G wireless networks raise serious concerns on privacy leakage. In HetNets, due to the high density of small cells, the association algorithm can reveal the location privacy of users. In a differential private algorithm is proposed to protect the location privacy. In the privacy in group communications is secured by the proposed protocol. In cryptographic mechanisms and schemes are proposed to provide secure and privacy-aware real-time video reporting service in vehicular networks.

- Availability: Availability is defined as the degree to which a service is accessible and usable to any legitimate users whenever and wherever it is requested. Availability evaluates how robust the system is when facing various attacks and it is a key performance metric in 5G. Availability attack is a typical active attack. One of the major attacks on availability is DoS attack, which can cause service access denial to legitimate users. Jamming or interference can disrupt the communication links between legitimate users by interfering the radio signals. With massive unsecured IoT nodes, 5G wireless networks face a big challenge on preventing jamming and DDoS attacks to ensure the availability service.

For the availability at PHY, DSSS and FHSS are two classical PLS solutions. DSSS was first applied to the military in 1940s. A pseudo noise spreading code is multiplied with the spectrum of the original data signal in DSSS. Without knowledge on the pseudo noise spreading code, a jammer needs a much higher power to disrupt the legitimate transmission. For FHSS, a signal is transmitted by rapidly switching among many frequency channels using a pseudorandom sequence generated by a key shared between transmitter and receiver. Dynamic spectrum is applied to D2D communications and cognitive radio paradigm to improve the SE in 5G. In FHSS can cause bad performance with the jamming attack. A pseudorandom time hopping spread spectrum is proposed to improve the performance on jamming probability, switching probability, and error probability. Resource allocation is adopted to improve the detection of the availability violation.

- Integrity: Although message authentication provides the corroboration of the source of the message, there is no protection provided against the duplication or modification of the message. 5G aims to provide connectivity anytime, anywhere, and anyhow, and to support applications closely related to human being daily life such as metering for the quality of the drinking water and scheduling of the transportation. The integrity of data is one of the key security requirements in certain applications.

Integrity prevents information from being modified or altered by active attacks from unauthorized entities. Data integrity can be violated by insider malicious attacks such as message injection or data modification. Since the insider attackers have valid identities, it is difficult to detect these attacks. In use cases such as smart meters in smart grid, data integrity service needs to be provided against manipulation. Compared to voice communications, data can be more easily attacked and modified. Integrity services can be provided by using mutual authentication, which can generate an integrity key. The integrity service of personal health information is required. Message integrity can be provided in the authentication schemes.

State-of-the-art in 5G Wireless Security

Many new PHY technologies in 5G wireless networks launched considerable research work in PLS. Most PLS research work are based on resource allocation. In a security-oriented resource allocation scheme is considered in ultra-dense networks (UDNs). The main resource dimensions mentioned are power allocation, relay selection, frequency allocation, time allocation, and beamforming. The open issues and future directions in PLS, including interference management, substitute for dedicated jammer, security over mobility management, and handing the heterogeneity. A case study for cross layer cooperation scheme in HetNet is presented when considering multiple users and SBSs in UDNs. For better understanding the PLS, two metrics used to evaluate the security performance are introduced as secrecy capacity and secrecy outage probability. The secrecy capacity C_s is defined as:

$$C_s = C_m - C_e,$$

where the C_m is the main channel capacity of the legitimate user, and the C_e is the channel capacity of the eavesdropper. The secrecy outage probability is defined as the instantaneous secrecy capacity is less than a target secrecy rate R_t, where $R_t > 0$, and:

$$P_{out}(R_s) = P(c_s < R_t).$$

Besides these two metrics, with the consumed power, in secrecy EE is defined as the ratio between the system achievable secrecy rate and the corresponding consumed power.

The new development and solutions in cryptography have mainly targeted at new applications. There have been development and proposed solutions on the security services including authentication, availability, confidentiality, and key management. Due to the escalated privacy concerns in 5G wireless networks, we further separate the confidentiality solutions into data confidentiality based and privacy based.

Authentication

Authentication is one of the most important security services in 5G wireless networks. In the legacy cellular networks, an authentication scheme is normally symmetric-key based. The implementation of the authentication scheme can deliver several security requirements. In the third generation (3G) cellular networks, the mutual authentication is implemented between a mobile station and the network. Following the authentication, a cipher key and an integrity key are generated to ensure both data confidentiality and integrity between the mobile station and the base station.

Due to the low latency requirement of 5G networks, authentication schemes are required to be more efficient in 5G than ever before. To leverage the advantages of SDN, in a fast authentication scheme in SDN is proposed, which uses weighed secure-context-information (SCI) transfer as a noncryptographic security technique to improve authentication efficiency during high frequent handovers in a HetNet in order to address the the latency requirement. Compared with the digital cryptographic authentication methods, the proposed method is hard to be totally compromised since it is based on the user-inherent physical layer attributes. There are more than one physical layer characteristics used in SCI to improve the authentication reliability for applications requiring a high level of security. The SDN enabled authentication model is shown in figure. The SDN controller implements an authentication model to monitor and predict the user location in order to prepare the relevant cells before the user arrival. This helps achieve seamless handover authentication. Physical layer attributes are used to provide unique fingerprints of the user and to simplify authentication procedure. Three kinds of fingerprints are used as the user-specific physical layer attributes. The validated original attributes are obtained after a full authentication. The observations are collected through constantly sampling multiple physical layer attributes from the received packets at the SDN controller. Both the original file and observation results contain the mean value of the attributes and variance of the chosen attributes. Then the mean attribute offset can be calculated based on the validated original attributes and observed attributes. If the attribute offset is less than a pre-determined threshold, the user equipment is considered legitimate. The detection probability is presented in the paper. To evaluate the performance of the proposed method, a SDN network model using priority queuing is proposed. The arriving traffic is modeled as a Pareto distribution. Authentication delay is compared among different network utilization scenarios. The proposed fast authentication protocol includes full authentication and weighted SCI transfer based fast authentication. As shown in figure after the first full authentication in one cell, it can be readily applied in other cells with MAC address verification, which only needs local processing. Moreover, full authentication can even be done without disrupting the user communication. A valid time duration parameter is used to flexibly adjust the secure level requirement. The simulation results compared the delay performance between the SDN enabled fast authentication and the conventional cryptographic authentication method. The SDN enabled fast authentication has a better delay performance owing to SDN flexibility and programmability in 5G networks.

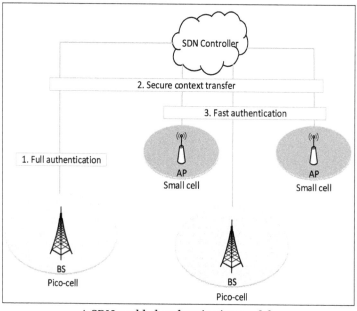

A SDN enabled authentication model.

To address the issues caused by the lack of a security infrastructure for D2D communications, in a securityscoring based on continuous authenticity is developed to evaluate and improve the security of D2D wireless systems. The principle of legitimacy patterns is proposed to implement continuous authenticity, which enables attack detection and system security scoring measurement. For the legitimacy pattern, a redundant sequence of bits is inserted into a packet to enable the attack detection. The simulation results show the feasibility of implementing the proposed security scoring using legitimacy patterns. Legitimacy patterns considering technical perspectives and human behaviors could improve the performance.

Combining the high security and utmost efficiency in bandwidth utilization and energy consumption in 5G, in a new cyclic redundancy check (CRC)-based message authentication which can detect any double-bit errors in a single message. The CRC codes based cryptographic hash functions are defined. A linear feedback shift register (LFSR) is used to efficiently implement the CRC encoding and decoding. The message authentication algorithm outputs an authentication tag based on a secret key and the message.

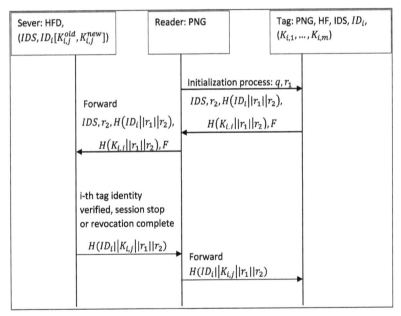

It is assumed that the adversary has the family of hash functions but not the particular polynomial g(x) and the pad s that are used to generate the authentication tag. The generator polynomial is changed periodically at the beginning of each session and pad s is changed for every message. The new family of cryptographic hash functions based on CRC codes with generator polynomials in g(x) = (1 + x)p(x) are introduced, where p(x) is a primitive polynomial. The proposed CRC retains most of the implementation simplicity of cryptographically non-secure CRCs. However, the applied LFSR requires re-programmable connections.

Radio frequency identification (RFID) has been widely applied and a single RFID tag can integrate multiple applications. Due to various limitations in low-cost RFID tags, the encryption algorithms and authentication mechanisms applied to RFID systems need to be very efficient. Thus simple and fast hash function are considered for the authentication mechanisms. Moreover, with multiple applications of single RFID, the revocation should be taken consideration into the authentication scheme. In the authors proposed a revocation method in the RFID secure authentication scheme in 5G use cases. A hash function and a random number are used to generate the corresponding module through a typical challenge-response mechanism. Figure shows the authentication process of the RFID secure application revocation scheme. The reader contains a pseudo-random number generator (PNG) and the sever holds a hash function and a database (HFD). The server establishes a tag record for each legitimate tag as (IDS, IDi) and a group of corresponding application records as $(K_{i,j}^{old}, K_{i,j}^{now})$. q is the authentication request generated by the reader. r_1 is the first random number generated by the PNG in reader. After receiving the authentication request, the tag generates the second random number r_2 and calculates two hash authentication messages M_1, M_2, and value of XOR authentication information $F = E \oplus K_{i,j}$, where E is the current value of the status flag information, which is used to determine whether to revoke or to certify the application. The security and complexity results are presented, which show that the proposed scheme has a higher level of security and the same level of complexity compared with existing ones.

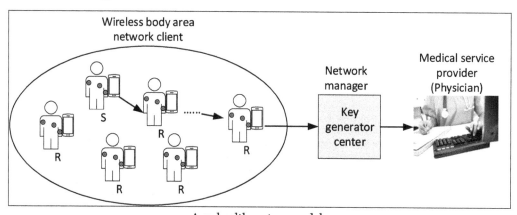

A m-health system model.

Considering the open nature of D2D communications between medical sensors and the high privacy requirements of the medical data, in by utilizing certificate-less generalized signcryption (CLGSC) technique, the authors proposed a light-weight and robust security-aware (LRSA) D2D-assist data transmission protocol in a m-health system. The m-health system is modeled in figure where S indicates the source node, and R represents the relay node. The anonymous and mutual authentication is implemented between the client and the physician in a wireless body area network to protect the privacy of both the data source and the intended destination. The signcryption of the message μs and encryption of its identity e_H^S are applied to the source client to authenticate the physician. A certificated-less signature algorithm is applied to the source client data before it is sent out. The source data identity can only be recovered by the intended physician who has the private key (x_H, z_H). The cipher text μs should be decrypted after the source identity is recovered with the right session key. Therefore, even the private key is leaked out, without the session key, the ciphertext is still safe. On the other hand, by verifying the signcryption μs, the physician can authenticate the source client. The relay nodes can verify the signature and then forward the data with their own signatures. The computational and communication overheads of the proposed CLGSC are compared with other four schemes. Simulation results show that the proposed CLGSC scheme has a lower computational overhead than the other four schemes.

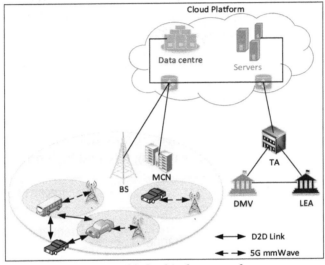

A 5G-enabled vehicular network.

Compared to IEEE 802.11p and the legacy cellular networks, 5G is a promising solution to provide real-time services for vehicular networks. However, the security and privacy need to be enhanced in order to ensure the safety of transportation. In a reliable, secure, and privacy-aware 5G vehicular network supporting real-time video services is presented. The system architecture is shown in Fig, which includes a mobile core network (MCN), a trusted authority (TA), a department of motor vehicles (DMV), and a law enforcement agency (LEA). D2D communications and mmWave techniques are adopted in the 5G vehicular communications. As shown in figure, HetNet is applied to expand network capacity and achieve high user data rates. The cloud platform provides massive storage and ubiquitous data access. The proposed cryptographic mechanisms include a pseudonymous authentication scheme, a public key encryption with keyword search, a ciphertextpolicy attribute-based encryption, and threshold schemes based on secret sharing. The pseudonymous authentication scheme with strong privacy preservation is applied to optimize the certification revocation list size, which is in a linear form with respect to the number of revoked vehicles so that certification verification overhead is the lowest. The authentication requirements include vehicle authentication and message integrity, where vehicle authentication allows the LEA and official vehicles to check the sender authenticity. The authentication is achieved by using a public-key-based digital signature that binds an encrypted traffic accident video to a pseudonym and to the real identity of the sender. The pseudonymous authentication technique can achieve the conditional anonymity and privacy of the sender.

Availability

Availability is a key metric to ensure the ultra-reliable communications in 5G. However, by emitting wireless noise signals randomly, a jammer can degrade the performance of the mobile users significantly and can even block the availability of services. Jamming is one of the typical mechanisms used by DoS attacks. Most of the anti-jamming schemes use the frequency-hopping technique, in which users hop over multiple channels to avoid the jamming attack and to ensure the availability of services.

In the authors proposed a secret adaptive frequency hopping scheme as a possible 5G technique against DoS based on a software defined radio platform. The proposed bit error rate (BER) estimator based on physical layer information is applied to decide frequency blacklisting under DoS attack. Since the frequency hopping technique requires that users have access to multiple channels, it may not work efficiently for dynamic spectrum access users due to the high switching rate and high probability of jamming.

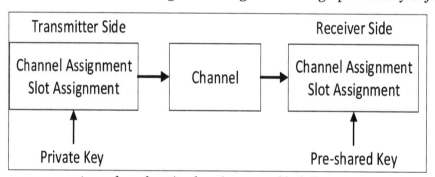

A pseudorandom time hopping system block diagram.

To reduce the switching rate and probability of jamming, in a pseudorandom time hopping anti-jamming scheme is proposed for cognitive users in 5G to countermeasure jamming attacks. The impact of spectrum dynamics on the performance of mobile cognitive users is modeled with the presence of a cognitive jammer with limited resources. The analytical solutions of jamming probability, switching rate, and error probability are presented. The jamming probability relates to delay performance and error probability. The jamming probability is low when the jammer lacks the access opportunities. Switching probability of time-hopping system outperforms the frequency-hopping system. With the same average symbol energy per joule, time-hopping has a lower error probability than frequency-hopping, and the performance gain saturates at a certain symbol energy level. The authors pointed out that the proposed time-hopping technique is a strong candidate for D2D links in 5G wireless networks due to its good EE and SE performance as well as its capability in providing jamming resilience with a small communication overhead. However, a pre-shared key is required for the time-hopping anti-jamming technique. The pseudorandom time hopping system block diagram is shown in figure. Both frequency hopping and time hopping require a pre-shared key to determine the hopping sequence.

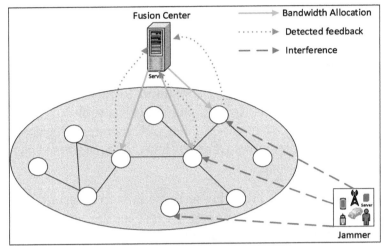

The resource allocation model.

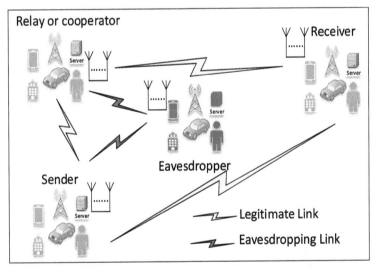

A general system model with eavesdropping attacks.

Considering the limited computational capabilities at certain nodes, in a fusion center is used to defend these nodes from a malicious radio jamming attack over 5G wireless network. A noncooperative Colonel Blotto game is formulated between the jammer and the fusion center as an exercise in strategic resource distribution. Figure shows the resource allocation model between fusion center and the malicious jammer. The jammer aims to jeopardize the network without getting detected by distributing its power among the nodes intelligently. On the other hand, the fusion center as a defender aims to detect such an attack by a decentralized detection scheme at a certain set of nodes. The fusion center can allocate more bits to these nodes for reporting the measured interference. A hierarchal degree is assigned to each node based on its betweenness centrality. Once the attack is detected, the fusion center will instruct the target node to increase its transmit power to maintain a proper SINR for normal communications. The simulation results show that error rate performance improves significantly with the fusion center having more bits to allocate among the nodes. The proposed resource allocation mechanism outperforms the mechanism that allocates the available bits in a random manner.

Data Confidentiality

Data confidentiality service is commonly required to tackle eavesdropping attacks. The general system model with eavesdropping attacks is shown in figure. The specific system models can be different in the number of transmitter/receiver/eavesdropper antennas and in the number of eavesdroppers/relays/cooperators. The relays or cooperators are optional in the system.

- Power Control: Power control for security aims to control the transmit power to ensure that the eavesdropper can not recover the signal. Based on the most simple eavesdropping attack model with a single eavesdropper armed with a single antenna, in the authors proposed a distributed algorithm to secure D2D communications in 5G, which allows two legitimate senders to select whether to cooperate or not and to adapt their optimal power allocation based on the selected cooperation framework. Figure shows a general system model with eavesdropping attacks. In the system model in the sender, relay or cooperator, receiver, and eavesdropper are named as Alice, John, Bob, and Eve, respectively. Each user has a single antenna. A shared bi-directional link is applied between Alice and John. The problem is formulated to maximize the achievable secrecy rates for both Alice and John as follows:

$$C_a = \max \left(R_{ajb} - R_{ae} \right),$$

$$s.t. P_j + P_{jb} \leq P_J,$$

$$C_j = \max(R_{jab} - R_{je}),$$

$$s.t. P_a + P_{ab} \leq P_A,$$

where C_a and C_j represent the secrecy rates of Alice and John respectively. R_{ajb} and R_{jab} are the achievable rates of Alice and John respectively with helping to relay data for each other. R_{ae} and R_{je} are the achievable rates of eavesdropper from Alice and from John respectively. Eq. $s.t. P_j + P_{jb} \leq P_J$; and Eq. $s.t. P_a + P_{ab} \leq P_A$, represent the transmit power limitation of the two legitimate senders. Two cooperation scenarios are considered, namely cooperation with relay and cooperation without relay. In the cooperation with relay scenario, Alice and John can help relay data of each other using the shared bi-directional link. In cooperation without relay, Alice and John coordinate their respective transmission power to maximize the secrecy rate of the other one. The optimization problem of noncooperation scenario is also presented for comparison. The distance between the legitimate transmitter and the eavesdropper is given a constraint to avoid distance attacks as the eavesdropper may have a better received signal quality on the transmitted message than the legitimate receiver. Simulation results show that achievable secrecy rates of Alice and John are improved by relaying data for each other. With the increase of distance between the transmitter and the receiver, the benefit from cooperation decreases and at some point non-cooperation could become more beneficial to the legitimate transmitter.

With no relay or cooperation, based only on power control and channel access, in a Stackelberg game framework for analyzing the achieved rate of cellular users and the secrecy rate of D2D users in 5G by using PLS. The system model includes one base station (BS), a number of cellular users, one D2D link, and one eavesdropper, as shown in figure. The utility function of cellular user achieved rates and D2D user secrecy rates are expressed as functions of channel information and transmission power:

$$u_{c,i} = \log_2(1 + SINR_{c,i}) + \alpha\beta P_D h_{dc},$$

$$u_d = [\log_2(1 + SINR_d) - \log_2(1 + SINR_e)] - \alpha P_D h_{dc},$$

where α is the price factor and is the scale factor. The first term in $u_{c,i}$ represents the data rate of the ith cellular user, and the second term compensates the interference from the D2D link, where PD is the transmit power of the D2D user and hdc is the channel gain from the D2D user to cellular users. The utility function of D2D user includes the secrecy data rate and the payment for the interference to cellular users. The game strategy of cellular users depends on the price factor α and game strategy of D2D user depends on the transmission power PD. The Stackelberg game is formed to maximize cellular utility function at the first stage and then the utility function of D2D user at the second stage.

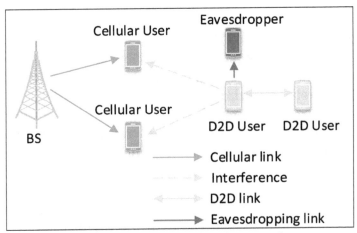

The system model with D2D link and an eavesdropper.

Power control is also one of the normally used mechanisms to improve the EE of the network. In the trade-off between PLS and EE of massive MIMO in an HetNet. An optimization model is presented to minimize the total power consumption of the network while satisfying the security level against eavesdroppers by assuming that the BS has imperfect channel knowledge on the eavesdroppers. The simulation results show that a highly dense network topology can be an effective solution to achieve high capacity, high cellular EE, and reliable and secure communication channels.

- Relay: Cooperation with relay can be used to help the sender to secure the signal transmission. In two relay selection protocols, namely optimal relay selection (ORS) and partial relay selection (PRS), are proposed to secure an energy harvesting relay system in 5G wireless networks. The system model is shown in figure, which consists of multiple relay nodes and assumes there is no direct link between sender and receiver. The power beacon is armed with multiple antennas, which can be used to strengthen the energy harvested. The ORS chooses the aiding relay to maximize the secrecy capacity of the system by assuming the source has full knowledge of channel state information (CSI) on each link. The PRS selects the helping relay based on partial CSI. The system includes a power beacon with multiple antennas, several relays, a destination node and an eavesdropper with a single antenna. Two energy harvesting scenarios that aim to maximize energy harvesting for source and selected relay are investigated. The analytical and asymptotic expressions of secrecy outage probability for both relay selections protocols are presented. The numerical results show that ORS can significantly enhance the security of the proposed system model and can achieve full secrecy diversity order while PRS can only achieve unit secrecy diversity order regardless of the energy harvest strategies. PRS that maximizes energy harvesting for relay strategy has a better secrecy performance than the one based on the maximizing energy harvesting for source. Moreover, the results show that the secrecy performance of the considered system is impacted significantly by the duration of energy harvest process.

To tackle the complexity issue of relay selection in 5G largescale secure two-way relay amplify-and-forward (TWR-AF) systems with massive relays and eavesdroppers, in the authors proposed a distributed relay selection criterion that does not require the information of sources SNR, channel estimation, or the knowledge of relay eavesdropper links. The proposed relay selection is done based on the received power of relays and knowledge of the average channel information between the source and the eavesdropper. The system model includes two source nodes, a number of legitimate relay nodes and multiple passive eavesdroppers. Each node has a single antenna. The cooperation of eavesdroppers is considered. In TWR-AF, the received signals from the two sources at the eavesdropper in each time slot are overlapped, where one source's signal acts as the jamming noise. The analytical results show that the number of eavesdroppers has a severe impact on the secrecy performance. The simulation results show that the performance of the proposed low-complexity criterion is very close to that of the optimal selection counterpart.

Considering eavesdroppers and relay with both single and multiple antennas, in the transmission design for secure relay communications in 5G networks is studied by assuming no knowledge on the number or the locations of eavesdroppers. The locations of eavesdroppers form a homogeneous Poisson Point Process. A randomize-and-forward relay strategy is proposed to secure multi-hop communications. Secrecy outage probability of the two-hop transmission is derived. A secrecy rate maximization problem is formulated with a secrecy outage probability constraint. It gives the optimal power allocation and codeword rate. Simulation results show that the secrecy outage probability can be improved by equipping each relay with multiple antennas. The secrecy throughput is enhanced and secure coverage is extended by appropriately using relaying strategies.

- Artificial Noise: Artificial noise can be introduced to secure the intended signal transmission. With the artificial-noiseaided multi-antenna secure transmission under a stochastic geometry framework, in the authors proposed an association policy that uses an access threshold for each user to associate with the BS so that the truncated average received signal power beyond the threshold is maximized and it can tackle randomly located eavesdroppers in a heterogeneous cellular network. The tractable expression of connection probability and secrecy probability for a randomly located legitimate user are investigated. Under the constraints of connection and secrecy probabilities, the network secrecy throughput and minimum secrecy throughput of each user are presented. Numerical results are presented to verify the analytical accuracy.

Assuming the sender is armed with multiple antennas, in an artificial noise transmission strategy is proposed to secure the transmission against an eavesdropper with a single antenna in millimeter wave systems. Millimeter wave channel is modeled with a ray cluster based spatial channel model. The sender has partial CSI knowledge on the eavesdropper. The proposed transmission strategy depends on directions of the destination and the propagation paths of the eavesdropper. The secrecy outage probability is used to analyze the transmission scheme. An optimization problem based on minimizing the secrecy outage probability with a secrecy rate constraint is presented. To solve the optimization problem, a closed-form optimal power allocation between the information signal and artificial noise is derived. The secrecy performance of the millimeter wave system is significantly influenced by the relationship between the propagation paths of destination and eavesdropper. The numerical results show that the secrecy outage is mostly occurred if the common paths are large or the eavesdropper is close to the transmitter.

To improve EE of the security method using artificial noise, in an optimization problem is formulated to maximize the secrecy EE by assuming imperfect CSI of eavesdropper at transmitter. The system is modeled with one legitimate transmitter with multiple antennas, and one legitimate receiver and one eavesdropper, each with a single antenna. Artificial noise is used at the transmitter. Resource allocation algorithms are used to solve the optimization problem with correlation between transmit antennas. With the combination of fractional programming and sequential convex optimization, the firstorder optimal solutions are computed with a polynomial complexity.

- Signal Processing: Besides the three methods above to provide data confidentiality, in the authors proposed an original symbol phase rotated (OSPR) secure transmission scheme to defend against eavesdroppers armed with unlimited number of antennas in a single cell. Perfect CSI and perfect channel estimation are assumed. The BS randomly rotates the phase of original symbols before they are sent to legitimate user terminals. The eavesdropper can not intercept signals, only the legitimate users are able to infer the correct phase rotations recover the original symbols. Symbol error rate of the eavesdropper is studied, which proves that the eavesdropper can not intercept the signal properly as long as the base station is equipped with a sufficient number of antennas.

Considering multiple eavesdroppers in the secure performance on a large-scale downlink system using non-orthogonal multiple access (NOMA). The system considered contains one BS, M NOMA users and eavesdroppers randomly deployed in an finite zone. A protected zone around the source node is adopted for enhancing the security of the random network. Channel statistics for legitimate receivers and eavesdroppers and secrecy outage probability are presented. User pair technique is adopted among the NOMA users. Analytical results show that the secrecy outage probability of NOMA pairs is determined by the NOMA users with poorer channel conditions. Simulation results show that secrecy outage probability decreases when the radius of the protected zone increases and secrecy outage probability can be improved by reducing the scope of the user zone as the path loss decreases.

In the authors proposed a dynamic coordinated multipoint transmission (CoMP) scheme for BS selection to enhance secure coverage. Considering co-channel interference and eavesdroppers, analysis of the secure coverage probability is presented. Both analytical and simulation results show that utilizing CoMP with a proper BS selection threshold the secure coverage performance can be improved, while secure coverage probability decreases with the excessive cooperation.

The proposed CoMP scheme has a better performance to resist more eavesdroppers than the no-CoMP scheme.

In massive MIMO is applied to HetNets to secure the data confidentiality in the presence of multiple eavesdroppers. The tractable upper bound expressions for the secrecy outage probability of HetNet users are derived, which show that massive MIMO can significantly improve the secrecy performance. The relationship between the density of picocell base station and the secrecy outage probability of the HetNet users.

- Cryptographic Methods: Besides the PLS solutions introduced above, cryptographic methods are also used for implementing data confidentiality by encrypting data with secret keys. Asymmetric cryptography can be applied to key distributions. To reduce the cost of encryption, symmetric cryptography is adopted for data encryption.

In a participating vehicle can send its random symmetric key, which is encrypted using TA's public key. The symmetric key is used to encrypt the message between TA, DMV, and participating vehicles. A one-time encryption key is also encrypted by a public key. The one-time encryption key is used to encrypt the video. In, an initial symmetric session key is negotiated between the client and a physician after they establish the client/server relationship. The symmetric key is then used for the data transmission between the client and the physician.

Key Management

Key management is the procedure or technique that supports the establishment and maintenance of keying relationships between authorized parties, where the keying relationship is the way common data is shared between communication entities. The common data can be public or secret keys, initialization values, and other non-secret parameters.

To provide flexible security, in+ three novel key exchange protocols, which have different levels of computational time, computational complexity, and security, for D2D communications are proposed based on the Diffie-Hellman (DH) scheme. Details of the key exchange schemes are shown in figure. The threat analysis of all three proposed protocols under common brute force and MITM attacks is presented. Performance study is provided for the proposed protocols to evaluate the confidentiality, integrity, authentication, and nonrepudiation of security services based on theoretical analysis. The analysis proves that the proposed protocols are feasible with reasonable communication overhead and computational time.

For D2D group use cases, in a group key management (GKM) mechanism to secure the exchanged D2D message during the discovery and communication phases is proposed. There are five security requirements in the proposed GKM, namely forward secrecy (users that have left the group should not have access to the future key), backward secrecy (new users joining the session should not have access to the old key), collusion freedom (fraudulent users could not deduce the current traffic encryption), key independence (keys in one group should not be able to discover keys in another group), and trust relationship (do not reveal the keys to any other part in the same domain or any part in a different domain). ID-based cryptography (IBC) scheme based on Elliptic Curve Cryptography (ECC) for securing multicast group communications is presented. The steps of the proposed protocol include secret key generation, elliptic curve digital signature algorithm, signature verification, group formation procedure, key generation, join process, and leave process. The master key and private key generations are based on IBC and ECC schemes. The overhead for communications, re-keying message, and key storage are assessed. The weakness of the IBC scheme and the ways of creating and using GKM are compared. The overall performance comparisons show that the proposed GKM has an enhancement in both the protocol complexity and security level compared with other works.

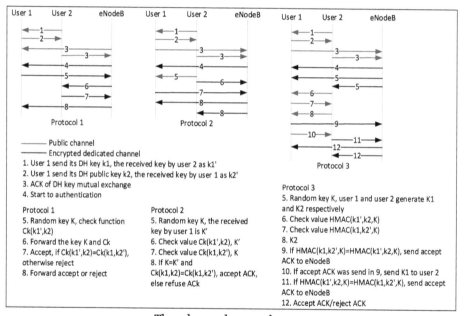

Three key exchange schemes.

ECC is also adopted for the proposed LRSA protocol. The network manager generates a partially private and partially public key for the client and the physician after the registration. And once the client and the physician establish the client/server relationship, an initial systematic session key can be set up for the data transmission.

Privacy

5G wireless networks raise serious concerns on privacy leakage when supporting more and more vertical industries such as m-health care and smart transportation. The data flows in 5G wireless networks carry extensive personal privacy information such as identity, position, and private contents. In some cases, privacy leakage may cause serious consequences. Depending on the privacy requirements of the applications, privacy protection is a big challenge in 5G wireless networks. There have already been research work considering location privacy and identity privacy.

Regarding location privacy, in to protect the location and preferences of users that can be revealed with associated algorithms in HetNets, a decentralized algorithm for access point selection is proposed based on a matching game framework, which is established to measure the preferences of mobile users and base stations with physical layer system parameters. Differentially private Gale-Shapley matching algorithm is developed based on differential privacy. Utilities of mobile users and access points are proposed based on packet success rate. Simulation results show that the differentially private algorithm can protect location privacy with a good quality of service based on utility of the mobile users. In a location-aware mobile intrusion prevention system (mIPS) architecture with privacy enhancement is proposed. The mIPS requirements, possible privacy leakage from managed security services.

In contextual privacy is defined as the privacy of data source and destination. The identity of the source client is encrypted by a pseudo identity of the source client with the public key of the physician using certificateless encryption mode. Meanwhile, the identity of the intended physician is also encrypted with the public key of the network manager. Through these two encryption steps, the contextual privacy can be achieved. For the proposed reporting service in privacy is an essential requirement to gain acceptance and participation of people. The identity and location information of a vehicle should be preserved against illegal tracing. Meanwhile, a reporting vehicle should be able to reveal its identity to the authorities for special circumstances. The pseudonymous authentication schemes are applied to achieve the conditional anonymity and privacy.

Security for Technologies Applied to 5G Wireless Network Systems

HetNet

HetNet is a promising technique to provide blanket wireless coverage and high throughput in 5G wireless networks. It is a multi-tier system in which nodes in different tier have different characteristics such as transmission power, coverage size, and radio access technologies. With the heterogeneous characteristics, HetNet achieves higher capacity, wider coverage and better performance in EE and SE. However, HetNet architecture, compared to single-tier cellular network, makes UE more vulnerable to eavesdropping. Moreover, with the high density of small cells in HetNet, traditional handover mechanisms could face significant performance issues due to too frequent handovers between different cells. The privacy issue in HetNet also faces a big challenge. Location information becomes more vulnerable due to the high density of small cells. The conventional association mechanism can disclose the location privacy information.

To tackle the eavesdropping attacks in HetNet, a secret mobile association policy is proposed based on the maximum truncated average received signal power (ARSP). The maximum ARSP should be higher than a pre-set access threshold in order for mobile to keep active. Otherwise, the mobile device remains idle. In the authors analyzed the user connection and secrecy probability of the artificialnoise-aided secure transmission with the proposed association policy, which is based on an access threshold. The secrecy throughput performance can be significantly enhanced with a proper access threshold used in the association policy.

For enhancing communication coverage in HetNet, coordinated multipoint transmission (CoMP) can be applied. However, CoMP can increase the risk of being eavesdropped for the legitimate users. In multiple BSs are selected to transmit the message. A dynamic BS selection scheme is proposed based on the secure coverage probability. Based on the theoretical and simulation results, the proper BS selection threshold for CoMP can improve the secure coverage performance.

Security-based resource management has been used to implement security in HetNet. In a case to improve the existing jamming and relaying mechanisms by proposing a cross-layer cooperation scheme with the aid of SBSs for protecting the confidentiality of macro cell user communications. The SBSs are motivated by monetary or resource bonus to become jammers to assist the secure communications under the constraints of the QoS of their own users.

Due to the high density of small cells, the knowledge of the cell an user is associated with can easily reveal the location information of that user. In the location privacy based on physical layer of association algorithms in 5G. A differential private Gale-Shapley algorithm is proposed to prevent the leakage of location information with certain QoS for users. The evaluation of the algorithm based on different privacy levels is presented with the influence on utility of users.

The intrusion detection based approach is considered as one way to provide secure communications. In intrusion detection techniques for mobile cloud computing in heterogeneous 5G are introduced. Several detection methodologies are studied as signature-based detection, anomaly-based detection, specification-based detection, stateful protocol analysis, hybrid intrusion detections with principles of these approaches. Traditional password-based authentication and biometric authentication are discussed for providing different levels of security.

D2D

In D2D communications, devices can communicate with each other without going through BSs. D2D communications enable efficient spectrum usage in 5G. Moreover, D2D communications can effectively offload traffic from BSs. However, the lack of a D2D security infrastructure makes the D2D communications less secure than the device to network communications. To improve the SE, dynamic spectrum access is usually adopted for D2D links, which can yield security threats such as jamming. The security issue becomes a major concern for direct radio communications and large-scale deployment of D2D groups.

Cooperation between D2D nodes is a popular way to secure the D2D communications against eavesdroppers. The legitimate transmitters with a common receiver can improve their reliable transmission rate through cooperation. In the authors proposed a cooperation scheme to secure D2D communications considering distance. Before the cooperation, devices can check the distance to test whether cooperation can improve the security of the communications. The distance constraints can be used to determine cooperation jointly, cooperation from one side, or no cooperation to maximize the achievable secrecy rate. With no specific requirements for the D2D communications, the proposed scheme can be applied to all D2D communications scenarios.

Besides cooperation, power control and channel access are also considered in securing D2D communications. In optimal power control and channel access of D2D link are proposed to maximize the achievable rate of cellular users and the physical layer secrecy rate of D2D links. The system model is shown in figure. The utility function of a single D2D user is modeled by considering PLS requirement and payment of interference from other D2D users. A Stackelberg game approach is used, where the price from cellular users are leaders and transmission power of D2D users are followers. The channel access problem of D2D links is discussed to maximize the achievable secrecy rate of D2D links and to minimize the interference to the cellular users.

To provide a measurement for security level, continuous authenticity with legitimacy patterns is proposed in to enable wireless security scoring. Security scoring based on probability of attack detection is applied to prevent, react, and detect attacks. The continuous legitimacy pattern is inserted into packets to authenticate the integrity and authenticity of transmissions.

Considering the assistance of the network, in key exchange protocols involved with the two D2D users and eNodeB are proposed. Two scenarios are considered. For the traffic offload scenario, D2D users are connected to the same eNodeB. For the social networking scenario, D2D link is required for the applications in each D2D user. Public channel and encrypted dedicated channel are applied to the process of key exchange. The eNodeB is involved in the initial key exchange and mutual authentication of the D2D users. Based on the role of eNodeB in the authentication process, three different key exchange protocols are proposed with different computational time and complexity.

The security algorithms and solutions for public cellular systems are not adapted to the short radio range D2D communications. The security issues in both proximity service discovery and communication phases for D2D communications are presented and addressed by proposing a group key management mechanism using IBC. Key distributions and key revocations are two problems in group key management (GKM). Five security requirements of GKM are defined and corresponding solutions are provided. A key graph is applied by dividing a group of members into subgroups to reduce the complexity of join process and leave process.

With the development of D2D technique, m-health applications are adopted to improve efficiency and quality of healthcare services. The security requirements for D2D communications used in m-health system are analyzed in. The protocol needs to secure the data that is not accessed by relays and to achieve mutual authentication between the source and the intended physician without interaction. It also requires light weight for mobile terminals with energy and storage constraints and needs to be robust enough to fight against threats as part of the keys can be exposed. A certificateless public key cryptography is applied to achieve the security requirements. The private key of a user is generated by both key generator center and the user, which makes the key generator center unaware about user's private key. Authentication is achieved by recognizing the public key. Security objectives of m-health network are defined as data confidentiality and integrity, mutual authentication, anonymity to anyone except intended physician, unlinkability, forward security and contextual privacy.

Massive MIMO

By utilizing a large number of antennas at BSs, massive MIMO can provide high EE and SE to support more users simultaneously. The large number of antennas at BSs can significantly improve the throughput, EE performance, and shift the most of signal processing and computation from user terminals to BSs. Moreover, massive MIMO can improve the security of communications. In PLS for a downlink K-tier HetNet system with multiple eavesdroppers. Each MBS is armed with large antenna arrays using linear zero-forcing beamforming. Both theoretical analysis and simulation results show that massive MIMO can significantly enhance the secrecy outage probability of the macrocell users.

However, eavesdropper can utilize massive MIMO to attack the legitimate communications. In the system model, massive MIMO at both BS and the eavesdropper. The antenna arrays of the eavesdropper are far more powerful. The OSPR approach is introduced. Theoretical and simulation analysis shows that the antenna number at the BS can significantly impact the security performance. With the number of antennas at the BS is sufficiently high, the massive MIMO eavesdropper fails to decode the majority of the original symbols while the legitimate users are able to recover the original symbols with only a limited number of antennas. Compared to other approaches involved in jamming, the proposed method has a higher EE.

SDN

By decoupling the control plane from the data plane, SDN enables centralized control of the network and brings promising methods to make the network management simpler, more programmable, and more elastic. Information can be shared between cells by using SDN. SDN can provide three key attributes, namely logically centralized intelligence, programmability, and abstraction so that scalability and flexibility of the network can be greatly improved and cost can be significantly reduced. A survey of software-defined mobile network (SDMN) and its related security problems are provided in the pros and cons of the SDN security. The pros of SDN security over traditional networks are shown in Table. Besides the pros of the SDN brought to 5G wireless networks, the new security issues caused by SDN are presented in Table, together with possible countermeasures.

Inthe limitations in present mobile networks. A SDMN architecture consisting of an application, control plane, and data plane is proposed, which integrates SDN, NFV and cloud computing. The security mechanisms in legacy cellular networks are presented with their limitations. The expected security advantages of SDMN are introduced. The security perspectives that can be improved through SDMN are listed. Besides the advantages as a service.

To address the threats in SDMN, in security attack vectors of SDN are presented.The network attacks by using attack graph. Analytic hierarchy process and technique are applied to calculate the node minimal effort for SDMN. A case study based on of SDMN, threat vectors for SDMN architecture are also presented. In the open issues of 5G security and trust based on NFV and SDN are elaborated. Corresponding security and trust frameworks are proposed, which use NFV Trust Platform as a service, security function as a service and trust functions MobileFow architecture is presented as an example to test the proposed vulnerability assessment mechanism.

Table: The pros of SDN security over traditional networks.

SDN characteristic	Attributed to	Security use
Global network view	Centralization Traffic statistics collection.	Network-wide intrusion detection. Detection of switch's malicious behavior Network forensics.
Self-healing mechanisms	Conditional rules Traffic statistics collection.	Reactive packet dropping Reactive packet redirection.
Increased control capabilities	Flow-based forwarding scheme.	Access control.

Table: New security issues that SDN networks are exposed to along with possible countermeasures.

Targeted level	Malicious behavior	Caused by	Possible countermeasures
Forwarding plane	Switch DoS	Limited forwarding table storage capacity Enormous number of flows Limited switchs buffering capacity.	Proactive rule caching, Rule aggregation, Increasing switchs buffering capacity, Decreasing switch-controller communication delay.
	Packet encryption and tunnel bypassing.	Invisible header fields.	Packet type classification based on traffic analysis.

Control plane	DDoS attack	Centralization Limited forwarding table storage capacity Enormous number of flows.	Controller replication Dynamic master controller assignment Efficient controller placement.
	Compromised controller attacks.	Centralization	Controller replication with diversity Efficient controller assignments.
Forwarding-control Link	MITM attacks	Communication message sent in clear Lack of authentication.	Encryption Use of digital signatures.
	Replay attacks	Communication message sent in clear Lack of time stamping.	Encryption Time stamp inclusion in encrypted messages.

Due to the high density of small cells in 5G, key management is difficult with user frequently joining and leaving the small cells. Moreover, speeding up the authentication process is essential to ensure the low latency requirement in 5G. In SDN is introduced into the system model to enable the coordination between different heterogeneous cells. A SDN controller is used to monitor and predict the user locations. The multiple physical layer characteristics are constantly sampled by the SDN controller to show the performance of the multiple SCI combination. The weighted SCI design and decision rules are proposed. The SDN mode uses the priority queuing and arriving traffic is modeled as a Pareto distribution. The latency performance of the SDN based authentication is shown to be better than the performance of traditional cryptographic methods based on different load situations. By pre-shared SCI over SDN, security framework can have a higher tolerance level to deal with failures of the network.

IoT

Due to the limited computation capability of IoT nodes, security services in 5G IoT devices need to be efficient and lightweight. Relaying has been considered as an effective mechanism in IoT networks to save the power of IoT nodes and also to extend the transmission coverage.

In a fusion center is used to protect IoT nodes with limited computation power from jammer. Each IoT node is equipped with a sensor to detect the interference. The betweenness centrality of each IoT node is taken consideration to measure the importance of the node over the network. The decentralized interference measurements are collected at the fusion center in regular intervals on a common control channel.

A certain level threshold and aggregated received interference power level are used to determine whether a jamming attack exists or not. The jammer knows the topology of the network and correspondingly allocates certain interference power to the IoT nodes to decrease their SINR. The fusion center can also allocate bandwidth to certain nodes to measure the interference level in order to detect the jammer attack. Therefore, a non-cooperative Colonel Blotto game between the jammer and the fusion center is formed as a resource distribution problem.

In the security of relay communications, IoT networks is introduced by considering power allocation and codeword rate design over two-hop transmission against randomly distributed eavesdroppers. The problem is formulated to maximize the secrecy rate. Both single- and multiple-antenna cases at relays and eavesdroppers are considered. It is shown that proper relay transmission can extend secure coverage and the increase of the number of antennas at relay nodes can improve the security level.

RFID is an automatic identification and data capture technology widely used in IoT networks. In, a RFID secure application revocation scheme is proposed to efficiently and securely use multi-application RFID and revoke applications in the tag. Based on theoretical analysis, the proposed scheme can achieve a higher level of security than other existing schemes.

Proposed 5G Wireless Security Architecture

5G wireless network architecture, based on which we further propose a corresponding security architecture. Identity management and flexible authentication based on the proposed 5G security architecture are analyzed. A handover procedure and signalling load analysis are studied to illustrate the advantages of the proposed 5G wireless security architecture.

5G Wireless Network Architecture

A 5G wireless network architecture. The illustrated general 5G wireless network architecture includes a user interface, a cloud-based heterogeneous radio access network, a next generation core, distributed edge cloud and central cloud. The cloud-based heterogeneous radio access network can combine virtualization, centralization and coordination techniques for efficient and flexible resource allocation. Based on different use cases, 3GPP classifies more than 70 different use cases into four different groups such as massive IoT, critical communications, network operation, and enhanced mobile broadband. In the cloud-based heterogeneous access network, besides the 3GPP access and non-3GPP access, other new radio technologies will be added for more efficient spectrum utilization. In the first stage of 5G, the legacy evolved packet core (EPC) will still be valid. Network slicing is applied to enable different parameter configurations for the next generation core according different use cases. New flexible service-oriented EPC based on network slicing, SDN, and NFV will be used in the next generation core as virtual EPC (VEPC) shown in the figure. The VEPC is composed of modularized network functions. Based on different use cases, the network functions applied to each VEPC can be various. In the VEPC, control plane and user plane are separated for flexibility and scalability of the next generation core. Edge cloud is distributed to improve the service quality. Central cloud can implement global data share and centralized control.

Compared with the legacy cellular networks, 5G wireless networks introduce some new perspectives and changes. (1) User equipment and services are not limited to regular mobile phone and regular voice and data services. Based on different use cases and requirements, user interfaces are classified into four different groups such as massive IoT, critical communications, network operation, and enhanced mobile broadband. Every use case can affect the radio access selection and VEPC functions. (2) In addition to 3GPP access and non-3GPP access in the cloud-based heterogeneous radio access network, 5G access network includes other new radios, which build the foundation of wireless standards for the next generation mobile networks for higher spectrum utilization. The new radios can support the performance and connectivity requirements of various use cases in 5G wireless networks. Moreover, there are many technologies applied to the access network to improve the network performance, such as massive MIMO, HetNet, and D2D communications. (3) The next generation core will be based on cloud using network slicing, SDN and NFV to handle different use cases. The flexible service-oriented VEPC will be applied. With network slicing, SDN and NFV, different network functions can be applied to the service-oriented VEPC for different use cases. The next generation core is expected to be access-independent. Separation of control and user plane is important to achieve an access-agnostic, flexible and scalable architecture. (4) Edge cloud is applied to 5G wireless network to improve the performance of the network, such as latency.

5G Wireless Security Architecture

Based on the illustrated 5G wireless network architecture, we propose a 5G wireless security architecture as shown in figure. With the new characteristics of the next generation core, a separation of data plane and control plane of VEPC is proposed, where the data plane can be programmable for its flexibility. The major network functions in the control plane of the next generation core are identified in TR 23.799, which are utilized in our proposed security architecture as follows:

- Access and mobility management function (AMF): The function is applied to manage access control and mobility, which is implemented in MME for legacy cellular network. This can be vary with different use cases. Mobility management function is not necessary for fixed access applications.

- Session management function (SMF): Based on network policy, this function can set up and manage sessions. For a single AMF, multiple SMF can be assigned to manage different sessions of a single user.

- Unified data management (UDM): UDM manages subscriber data and profiles (such as authentication data of users) for both fixed and mobile access in the next generation core.

- Policy control function (PCF): This function provides roaming and mobility management, quality of service, and network slicing. AMF and SMF are controlled by PCF. Differentiated security can be provided with PCF.

AMF and SMF are integrated in the legacy cellular networks as MME. The separation of AMF and SMF can support a more flexible and scalable architecture. In the network function based control plane, different network functions can be applied to different use cases.

Similar to the legacy cellular networks, four security domains are defined in figure as A, B, C, D. The details of these security domains are introduced as follows.

Network access security (A): The set of security features that provide the user interface to access the next generation core securely and protect against various attacks on the radio access link. The new physical layer technologies applied to the radio access network including massive MIMO, HetNet, D2D communications and mmWave bring new challenges and opportunities in network access security. This level has security mechanisms such as confidentiality and integrity protection between the user interface and radio access network. Current researches on network access security focus on providing user identity and location confidentiality, user data and signaling data confidentiality, and entity authentication.

Network domain security (B): The set of security features that protect against attacks in the wire line networks and enable different entities and functions to exchange signaling data and user data in a secure manner. As we can see in figure, this level security exists between access network and next generation core, control plane and user plane. Since new technologies such as cloud technique, network slicing and NFV are applied to 5G core and radio access network, new vulnerabilities in this level need to be addressed. However, with the separation of control plane and user plane, the amount of signaling data will be significantly reduced. The network function based control plane also reduces the required signaling overhead for data synchronization. Entity authentication, data confidentiality and data integrity are the main security services in this level. With the independent characteristics of access technologies of AMF, the network domain security performance can be simplified and improved.

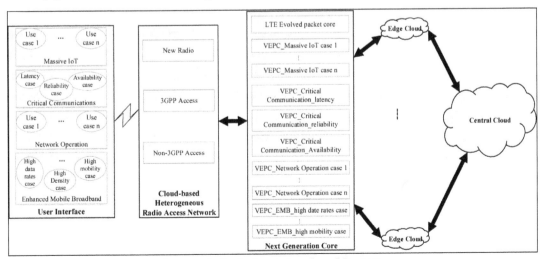

A general 5G wireless network architecture.

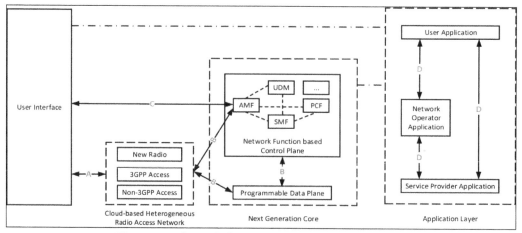

The proposed 5G wireless network security architecture.

User domain security (C): The set of security features that provide mutual authentication between the user interface and the next generation core before the control plane access to the user interface. Authentication is the main focus in this level. Based on the use case, the authentication may be needed for more than two parties. For example, the authentication can be required between user and network operator as well as between user and service provider. Moreover, different service providers may need to authenticate each other to share the same user identity management. Compared to the device-based identity management in legacy cellular networks, new identity management methods are needed to improve the security performance.

Application domain security (D): The set of security features that ensure the security message exchange between applications on the interfaces, between user interface and service provider, as well as between user and network operator.

5G Wireless Security Services

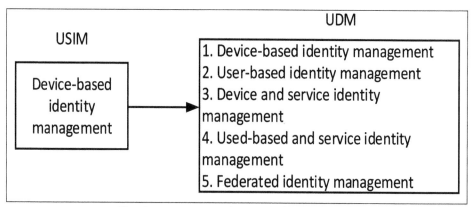

Identity management in 5G wireless networks.

Identity management: In the legacy cellular networks, the identity management relies on the universal subscriber identity module (USIM) cards. However, in 5G wireless networks, there are many equipment such as smart home devices, sensors and vehicles that are supported without USIM card. As shown in figure, UDM will handle the identity management based on cloud. Moreover, anonymity service is required in many use cases in 5G wireless networks. Therefore, the identity management will be different in 5G wireless networks compared with that in the legacy cellular networks. New identity management is required.

With the massive connected devices and applications, efficiently managing massive identities is significantly important to ensure the service performance. In the legacy cellular networks, the identity management is device-based. For a certain new use case such as smart home, one user can have multiple devices needed to access the network and services. User-based identity management will be more efficient to let the user determine what devices are allowed to access the network and services. One user may have multiple device identities. Except only considering the device identity, service identity can be added with device identity as device and service identity management. The device identity is unique and service identity can be assigned by service providers in certain session. With service identity, revocation process will be simplified.

Moreover, for the trusted service providers, federated identity management can be applied to simplify the identity management and also improve the user experience. The identity management in 5G wireless networks is not unified for all use cases. Based on the characteristics of the use case, different identity management can be applied as shown in the figure.

Flexible authentication: in the legacy cellular networks, mutual authentication is applied between a user and the network. However, the authentication between a user and the services provider is not implemented by the network. In 5G wireless network systems, some use cases may require both the service provider and network provider to carry out authentication with the users. In the legacy cellular networks, for 3GPP access, the AKA is applied between a user equipment and a mobile management entity. For non-3GPP access, AKA is applied between a user equipment and an authentication authorization and accounting (AAA) server. Full authentication is required once a user changes its access technology. Based on our proposed security architecture, AMF can handle the authentication independent of the access technologies. In other words, a full authentication is not required when a user changes its access technology. Moreover, based on PCF, AMF can perform different authentication schemes for different service requirements.

Flexible authentication is required in 5G wireless networks to ensure the security while satisfying the quality of services requirements. The input and output of the authentication mechanism selection are shown in figure. The input information can be included in PCF, which can control AMF to perform the authentication procedure.

Handover Procedure and Signaling Load Analysis

Analysis on handover procedure and signaling load are presented based on the proposed security architecture for a HetNet with different access technologies including 5G new radio, 3GPP access and Non-3GPP access. The system model is shown in figure, where a user A currently associates with 3GPP access point MBS. Assume that SBSs have different access technologies compared with MBS. When user A is moving, it may need to connect with a new radio access point (NRAP), in which case handover is needed in the legacy cellular networks. In our proposed security architecture, AMF is independent from different access technologies. User A can connect with the same AMF through different access technologies. The first time user A associates with an access point, a general authentication procedure is needed. Assume that the same authentication scheme is applied to the proposed 5G wireless network security architecture and the legacy security architecture. The authentication of first time access to the network for user A based on different security architectures is shown in figure. Since AMF and UDM are both in the control plane, the cost for information exchange between AMF and UDM is less than that between different entities such as MME and HSS. Based on the legacy security architecture, the authentication vector is generated at HSS and is then transmitted to MME. However, in our proposed security architecture, authentication vector can be generated at AMF to reduce the overhead of communications and to reduce the risk to expose the KASME and XRES. With the flexibility of network functions, AMF and UDM can be widely distributed to handle the authentication of a massive number of user devices. Nevertheless, due to the coupled control plane and user plane, MME and HSS have limited scalability.

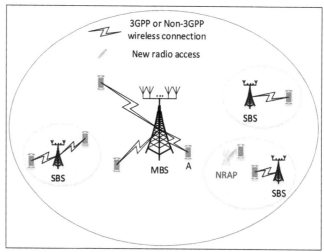

A two-tier HetNet model.

Once user A changes its access point using another access technology in legacy cellular networks, the same authentication as shown in figure is needed for each handover, which not only increases latency and communication overhead but also leads to possible connection outage. However, based on the proposed security architecture, no authentication will be needed by switching to different SMF for a new session and a new IP address allocation.

The handover based on the proposed 5G wireless security architecture is presented in figure. The data update from SMF includes the new session key and new IP address from the new access point. The communication latency between AMF and SMF can be neglected compared to the communication latency from MME to HSS. Moreover, the signaling overhead based on the 5G wireless security architecture is much lower because of the separation of control plane and user plane as shown in figure. To satisfy certain latency requirement, the number of gateway nodes needs to be increased by a factor of 20 to 30 times of the current number . The separation of control and user plane of gateway can also facilitate distributed gateway deployment. Therefore, for the new core network based on control and user plane separation, the signaling load can be significantly reduced.

Challenges and Future Directions for 5G Wireless Security

The challenges and future directions for 5G security research and development. Part of the security solutions used in 4G will be evolved into 5G. However, with extensive use cases and various integrated technologies applied to 5G, security services in 5G face many challenges in order to address 5G advanced features. Several perspectives of the challenges and corresponding future directions are discussed as follows.

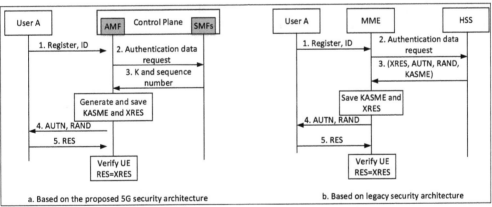

Authentication based on different security architecture.

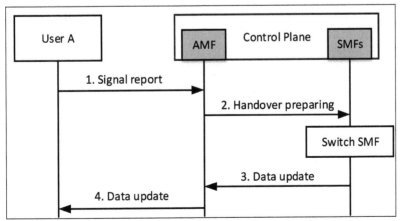

A handover procedure for access technologies change.

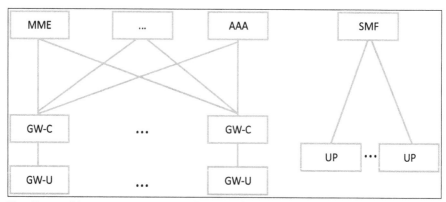

Signaling architecture comparison of legacy cellular network and 5G cellular network.

New Trust Models

With the advanced services offered by 5G wireless networks, not only new types of functions are provided to people and society, but also new services are applied to vertical industries, such as smart grid, smart home, vehicular networks and m-health networks, etc. In the legacy cellular networks, user terminals, home, and serving networks are considered in the trust model. The trust models vary among different use cases which can involve new actors in 5G wireless networks. The authentication may need to be implemented between various actors with multiple trust levels.

There have been research work on trust models for different use cases. In the authors proposed a system model to facilitate secure data transmission over 5G wireless networks for vehicular communications. DMV, TA, LEA, and vehicles are included in the proposed system model. The trust model between them is more complex than the trust model in the legacy cellular networks. With the massive number of devices over 5G wireless networks, new trust models are needed to improve the performance of security services such as IoT user cases authentication. However, it lacks a trust model between devices and fusion center. For some applications, there are various types of devices connected to the same network, some of which may be used only to gather data and some of which may be used only to access internet. The trust requirements of different devices should be different. For different security demands, the corresponding trust model may have different security requirements. As an example, a high security level demand may require both password and biometric authentication simultaneously. In a mhealth network, in the trust model between client, network management and physician based on the privacy requirements.

In summary, various new trust models for new applications in 5G are needed. These new trust models will affect the security services.

New Security Attack Models

Based on the recent research activities on PLS, the most used attack model consists of a single eavesdropper armed with a single antenna. However, the number of eavesdroppers can be high in 5G wireless networks. Moreover, eavesdroppers can be armed with massive MIMO technology . In practical scenarios, there may exist different types of attacks. By only considering one kind attack, the cooperation of jammer or eavesdroppers are not considered in PLS, which can make the security in PHY more complex. Although increasing the transmission power of the sender can fight against jamming attack, it may also increase the risk of eavesdropping attacks.

Moreover, with the new service delivery model applied to SDN and NFV, there are more vulnerable points exposed. Decoupling software from hardware makes the security of software no longer depending on the specific security attributes of the hardware platform. Therefore, the demands on strong isolation for virtualization are ever increasing. Network slicing is introduced to provide the isolated security. In an effective vulnerability assessment mechanism is proposed for SDN based mobile networks using attack graph algorithm. A comprehensive security attack vector map of SDN is presented.

The various new attack models in 5G wireless networks based on the new technologies and delivery models make the security implementation harder than in the legacy cellular networks. However, there has been limited work on the new security attack models and corresponding solutions.

Privacy Protection

With data involved in various new applications in 5G, huge volume of sensitive data are being transmitted through the 5G wireless networks. 5G wireless networks raise serious concerns on privacy leakage due to the open network platforms. The protection of the privacy is an important requirement for implementing different applications. The privacy protection in different use cases can vary based on the security requirements, such as location privacy, identity privacy. For example, in, to secure the privacy of patients, the proposed protocol provides security of data access and mutual authentication between patients and physician. The location privacy also draws great attention. In a differential private association algorithm is proposed to secure the location information due to the vulnerable location leakage in HetNets. For vehicular communications, in the privacy protection is considered as protection of the identity of a vehicle and the video contents. In order to offer differentiated quality of privacy protection, the type of service offered to a user needs to be sensed. However, the service type sensing may also have a chance to leak user privacy .

The privacy protection is mostly implemented by encryption mechanisms currently. With the massive data, encryption and decryption may violate other service requirements of 5G, such as latency and efficiency. To efficiently protect privacy is a big challenge, especially when facing the powerful data analysis methods such as machine learning. However, data analysis can also be used as a mechanism to help implement the privacy protection intelligently. For example, before the data transmission, data analysis can be applied to find out several highly sensitive dimensions to reduce the encryption cost with privacy protection. For the identity privacy, new identity management should be considered instead of using only device-based identity management. Location privacy can be enhanced if multiple association mechanisms are applied to different use cases. Adding all this together makes it more challenging to provide satisfactory privacy protection in 5G wireless networks.

Flexibility and Efficiency

To address different security requirements for different applications and dynamic configurations of the 5G architecture based on virtualization, the security mechanisms must be flexible. The security setup must be customized and optimized to support each specific application instead of an approach fitting all. Therefore, for each security service, different security levels need to be considered for different scenarios. If differentiated security is offered, a flexible security architecture is needed. In our proposed security architecture, network functions in the control plane are various depending on the use cases. AMF and SMF provide flexible security mechanisms based on the requirements of PCF. Therefore, the flexibility is not only required in security architecture but also in security mechanisms.

Besides the flexibility of security architecture and mechanisms, efficiency of security is another key requirement in 5G wireless networks to ensure both the latency requirement and EE. One of the potential security requirements is to minimize the security-related signaling overhead to ensure the efficiency. The latency can be reduced by reducing the overhead of security load. Since EE and latency performances of 5G wireless networks are expected to be improved compared to the legacy wireless networks, the security efficiency should be ensured to secure the performances of 5G wireless networks. Based on the proposed security architecture, the separation of control plane and user plane and network functions inside the control plane reduce the signaling overhead. For the IoT applications, the nodes normally have limited computation capability and battery power, efficient security mechanisms are required. Moreover, distributed authentication nodes need to support the fast network access for massive number of devices. For the vehicular communications sensitive to latency, lightweight and efficient security solutions are desirable. Moving the control plane closer to the edge of the core network can also reduce the communication latency. Therefore, to improve the efficiency of 5G wireless networks, both security architecture and security mechanisms need to be improved.

Unified Security Management

Although there are different services, access technologies and devices over 5G wireless networks, a security framework with a common and essential set of security features such as access authentication and confidentiality protection is needed . The basic features of these security services may be similar to those in the legacy cellular networks.

However, there are many new perspectives of these security features in 5G wireless networks, such as the security management across heterogeneous access and security management for a large number of devices. Flexible authentication and the handover between different access technologies based on the proposed security architecture, security management across heterogeneous access need to be defined to offer flexibility for all access technologies. Also, for a large number of devices, such as IoT applications, security management of burst access behavior need to be studied in order to support the efficient access authentication.

Mobile Communication Standards and Protocols

Mobile communication standards and protocols make use of multiplexing for sending information. GSM architecture, code division multiple access, general packet radio service, next generation network, etc. are some of the concepts within it. This chapter closely examines these key concepts of mobile communication protocols to provide an extensive understanding of the subject.

Any device that does not need to remain at one place to carry out its functions is a mobile device. So laptops, smartphones and personal digital assistants are some examples of mobile devices. Due to their portable nature, mobile devices connect to networks wirelessly. Mobile devices typically use radio waves to communicate with other devices and networks. Here we will discuss the protocols used to carry out mobile communication.

Mobile communication protocols use multiplexing to send information. Multiplexing is a method to combine multiple digital or analog signals into one signal over the data channel. This ensures optimum utilization of expensive resource and time. At the destination these signals are de-multiplexed to recover individual signals.

These are the types of multiplexing options available to communication channels:

- FDM (Frequency Division Multiplexing): Here each user is assigned a different frequency from the complete spectrum. All the frequencies can then simultaneously travel on the data channel.

- TDM (Time Division Multiplexing): A single radio frequency is divided into multiple slots and each slot is assigned to a different user. So multiple users can be supported simultaneously.

- CDMA (Code Division Multiplexing): Here several users share the same frequency spectrum simultaneously. They are differentiated by assigning unique codes to them. The receiver has the unique key to identify the individual calls.

Global System for Mobile Communication

GSM is a mobile communication modem; it is stands for global system for mobile communication (GSM). The idea of GSM was developed at Bell Laboratories in 1970. It is widely used mobile communication system in the world. GSM is an open and digital cellular technology used for transmitting mobile voice and data services operates at the 850MHz, 900MHz, 1800MHz and 1900MHz frequency bands.

GSM system was developed as a digital system using time division multiple access (TDMA) technique for communication purpose. A GSM digitizes and reduces the data, then sends it down through a channel with two different streams of client data, each in its own particular time slot. The digital system has an ability to carry 64 kbps to 120 Mbps of data rates.

GSM Modem.

There are various cell sizes in a GSM system such as macro, micro, pico and umbrella cells. Each cell varies as per the implementation domain. There are five different cell sizes in a GSM network macro, micro, pico and umbrella cells. The coverage area of each cell varies according to the implementation environment.

GSM Architecture

The GSM network architecture provided a simple and yet effective architecture to provide the services needed for a 2G cellular system.

There were four main elements to the overall GSM network architecture and these could often be further split. Elements like the base station controller, MSC, AuC, HLR, VLR and the like are brought together to form the overall system.

The 2G GSM network architecture, although now superseded gives an excellent introduction into some of the basic capabilities required to set up a mobile phone network and how all the entities operate together.

A base station antenna carrying 2G GSM signals.

GSM Network Architecture Elements

In order that the GSM system operates together as a complete system, the overall network architecture brings together a series of data network identities, each with several elements.

The GSM network architecture is defined in the GSM specifications and it can be grouped into four main areas:

- Network and Switching Subsystem (NSS).

- Base-Station Subsystem (BSS).

- Mobile station (MS).

- Operation and Support Subsystem (OSS).

The different elements of the GSM network operate together and the user is not aware of the different entities within the system.

As the GSM network is defined but he specifications and standards, it enables the system to operate reliably together regardless of the supplier of the different elements.

A basic diagram of the overall system architecture for 2G GSM with these four major elements is shown below:

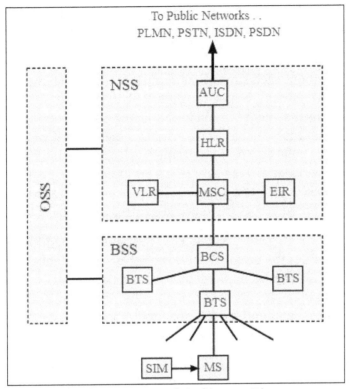

Simplified GSM Network Architecture Diagram.

Within this diagram the different network areas can be seen - they are grouped into the four areas that provide different functionality, but all operate to enable reliable mobile communications to be achieved.

The overall network architecture provided to be very successful and was developed further to enable 2G evolution to carry data and then with further evolutions to allow 3G to be established.

Network Switching Subsystem (NSS)

The GSM system architecture contains a variety of different elements, and is often termed the core network. It is essentially a data network with a various entities that provide the main control and interfacing for the whole mobile network. The major elements within the core network include:

- Mobile Services Switching Centre (MSC): The main element within the core network area of the overall GSM network architecture is the Mobile switching

Services Centre (MSC). The MSC acts like a normal switching node within a PSTN or ISDN, but also provides additional functionality to enable the requirements of a mobile user to be supported. These include registration, authentication, call location, inter-MSC handovers and call routing to a mobile subscriber. It also provides an interface to the PSTN so that calls can be routed from the mobile network to a phone connected to a landline. Interfaces to other MSCs are provided to enable calls to be made to mobiles on different networks.

- Home Location Register (HLR): This database contains all the administrative information about each subscriber along with their last known location. In this way, the GSM network is able to route calls to the relevant base station for the MS. When a user switches on their phone, the phone registers with the network and from this it is possible to determine which BTS it communicates with so that incoming calls can be routed appropriately. Even when the phone is not active (but switched on) it re-registers periodically to ensure that the network (HLR) is aware of its latest position. There is one HLR per network, although it may be distributed across various sub-centres to for operational reasons.

- Visitor Location Register (VLR): This contains selected information from the HLR that enables the selected services for the individual subscriber to be provided. The VLR can be implemented as a separate entity, but it is commonly realised as an integral part of the MSC, rather than a separate entity. In this way access is made faster and more convenient.

- Equipment Identity Register (EIR): The EIR is the entity that decides whether a given mobile equipment may be allowed onto the network. Each mobile equipment has a number known as the International Mobile Equipment Identity. This number, is installed in the equipment and is checked by the network during registration. Dependent upon the information held in the EIR, the mobile may be allocated one of three states - allowed onto the network, barred access, or monitored in case its problems.

- Authentication Centre (AuC): The AuC is a protected database that contains the secret key also contained in the user's SIM card. It is used for authentication and for ciphering on the radio channel.

- Gateway Mobile Switching Centre (GMSC): The GMSC is the point to which a ME terminating call is initially routed, without any knowledge of the MS's location. The GMSC is thus in charge of obtaining the MSRN (Mobile Station Roaming Number) from the HLR based on the MSISDN (Mobile Station ISDN number, the "directory number" of a MS) and routing the call to the correct visited MSC. The "MSC" part of the term GMSC is misleading, since the gateway operation does not require any linking to an MSC.

- SMS Gateway (SMS-G): The SMS-G or SMS gateway is the term that is used to collectively describe the two Short Message Services Gateways defined in the GSM standards. The two gateways handle messages directed in different directions. The SMS-GMSC (Short Message Service Gateway Mobile Switching Centre) is for short messages being sent to an ME. The SMS-IWMSC (Short Message Service Inter-Working Mobile Switching Centre) is used for short messages originated with a mobile on that network. The SMS-GMSC role is similar to that of the GMSC, whereas the SMS-IWMSC provides a fixed access point to the Short Message Service Centre.

These entities were the main ones used within the GSM network. They were typically co-located, but often the overall core network was distributed around the country where the network was located. This gave some resilience in case of failure.

Although the GSM system was essential a voice system, the core network was a data network as all signals were handled digitally.

Base Station Subsystem (BSS)

The Base Station Subsystem (BSS) section of the 2G GSM network architecture that is fundamentally associated with communicating with the mobiles on the network.

It consists of two elements:

- Base Transceiver Station (BTS): The BTS used in a GSM network comprises the radio transmitter receivers, and their associated antennas that transmit and receive to directly communicate with the mobiles. The BTS is the defining element for each cell. The BTS communicates with the mobiles and the interface between the two is known as the Um interface with its associated protocols.

- Base Station Controller (BSC): The BSC forms the next stage back into the GSM network. It controls a group of BTSs, and is often co-located with one of the BTSs in its group. It manages the radio resources and controls items such as handover within the group of BTSs, allocates channels and the like. It communicates with the BTSs over what is termed the Abis interface.

The base station subsystem element of the GSM network utilised the radio access technology to enable a number of users to access the system concurrently. Each channel supported up to eight users and by enabling a base station to have several channels, a large number of subscribers could be accommodated by each base station.

Base stations are carefully located by the network provider to enable complete coverage of an area. The area being covered bay a base station often being referred to as a cell.

As it is not possible to prevent overlap of the signals into the adjacent cells, channels used in one cell are not used in the next. In this way interference which would reduce call quality is reduced whilst still maintaining sufficient frequency re-use.

It is important to have the different BTSs linked with the BSS and the BSSs linked back to the core network.

A variety of technologies were used to achieve this. As data rates used within he GSM network were relatively low, E1 or T1 lines were often used, especially for linking the BSS back to the core network.

As more data was required with increasing usage of the GSM network, and also as other cellular technologies like 3G became more widespread, many links used carrier grade Ethernet.

Often remote BTSs were linked using small microwave links as this could reduce the need for the installation of specific lines if none were available. As base stations often needed to be located to provide good coverage rather than in areas where lines could be installed, the microwave link option provided an attractive method for providing a data link for the network.

Mobile Station

Mobile stations (MS), mobile equipment (ME) or as they are most widely known, cell or mobile phones are the section of a GSM cellular network that the user sees and operates. In recent years their size has fallen dramatically while the level of functionality has greatly increased. A further advantage is that the time between charges has significantly increased.

There are a number of elements to the cell phone, although the two main elements are the main hardware and the SIM.

The hardware itself contains the main elements of the mobile phone including the display, case, battery, and the electronics used to generate the signal, and process the data receiver and to be transmitted.

The mobile station, or ME also contains a number known as the International Mobile Equipment Identity (IMEI). This is installed in the phone at manufacture and "cannot" be changed. It is accessed by the network during registration to check whether the equipment has been reported as stolen.

The SIM or Subscriber Identity Module contains the information that provides the identity of the user to the network. It contains are variety of information including a number known as the International Mobile Subscriber Identity (IMSI). As this is included in the SIM, and it means that by moving the SIM card from one mobile to another, the user could easily change mobiles. The ease of changing mobiles whilst keeping the same number meant that people would regularly upgrade, thereby creating a further revenue stream for network providers and helping to increase the overall financial success of GSM.

Operation and Support Subsystem (OSS)

The OSS or operation support subsystem is an element within the overall GSM network architecture that is connected to components of the NSS and the BSC. It is used to control and monitor the overall GSM network and it is also used to control the traffic load of the BSS. It must be noted that as the number of BS increases with the scaling of the subscriber population some of the maintenance tasks are transferred to the BTS, allowing savings in the cost of ownership of the system.

The 2G GSM network architecture follows a logical method of operation. It is far simpler than current mobile phone network architectures which use software defined entities to enable very flexible operation. However the 2G GSM architecture does show the voice and operational basic functions that are needed and how they fit together. As the GSM system was all digital, the network was a data network.

Authentication Center (AuC)

The authentication center (AuC) is a key component of a global system for mobile communications (GSM) home locator register (HLR). The AuC validates any security information management (SIM) card attempting network connection when a phone has a live network signal.

The AuC provides security to ensure that third parties are unable to use network subscriber services.

Each network SIM card is assigned an individual authentication key (Ki). A matching Ki is contained in the AuC. The SIM and the AuC store the Kiin an unreadable format. The Ki even remains hidden from the SIM card owner to protect network operators from fraud, such as SIM cloning, through enabling user identity verification and ensuring call confidentiality.

During the authentication process, the Ki is used with the subscriber's international mobile subscriber identity (IMSI) number. Network and service subscriber validity is determined by successful authentication.

The authentication process begins when a subscriber requests a network signal. A randomly selected key is generated that encrypts all wireless communication between the mobile device and the core network. The encryption algorithm is known as A3.

The encrypted randomly chosen number (RAND) using the Ki must match the stored number in the AuC and the SIM card. The entire process is completed during a wireless connection. If the numbers do not match, the authentication is invalidated as a failed function request.

GSM Specification

The requirements for different Personal Communication Services (PCS) systems differ for each PCS network. Vital characteristics of the GSM specification are listed below:

Modulation

Modulation is the process of transforming the input data into a suitable format for the transmission medium. The transmitted data is demodulated back to its original form at the receiving end. The GSM uses Gaussian Minimum Shift Keying (GMSK) modulation method.

Access Methods

Radio spectrum being a limited resource that is consumed and divided among all the users, GSM devised a combination of TDMA/FDMA as the method to divide the bandwidth among the users. In this process, the FDMA part divides the frequency of the total 25 MHz bandwidth into 124 carrier frequencies of 200 kHz bandwidth.

Each BS is assigned with one or multiple frequencies, and each of this frequency is divided into eight timeslots using a TDMA scheme. Each of these slots are used for both transmission as well as reception of data. These slots are separated by time so that a mobile unit doesn't transmit and receive data at the same time.

Transmission Rate

The total symbol rate for GSM at 1 bit per symbol in GMSK produces 270.833 K symbols/second. The gross transmission rate of a timeslot is 22.8 Kbps.

GSM is a digital system with an over-the-air bit rate of 270 kbps.

Frequency Band

The uplink frequency range specified for GSM is 933 - 960 MHz (basic 900 MHz band only). The downlink frequency band 890 - 915 MHz (basic 900 MHz band only).

Channel Spacing

Channel spacing indicates the spacing between adjacent carrier frequencies. For GSM, it is 200 kHz.

Speech Coding

For speech coding or processing, GSM uses Linear Predictive Coding (LPC). This tool compresses the bit rate and gives an estimate of the speech parameters. When the audio signal passes through a filter, it mimics the vocal tract. Here, the speech is encoded at 13 kbps.

Duplex Distance

Duplex distance is the space between the uplink and downlink frequencies. The duplex distance for GSM is 80 MHz, where each channel has two frequencies that are 80 MHz apart.

Misc

- Frame duration : 4.615 mS.

- Duplex Technique : Frequency Division Duplexing (FDD) access mode previously known as WCDMA.

- Speech channels per RF channel : 8.

GSM - Addresses and Identifiers

GSM treats the users and the equipment in different ways. Phone numbers, subscribers, and equipment identifiers are some of the known ones. There are many other identifiers that have been well-defined, which are required for the subscriber's mobility management and for addressing the remaining network elements. Vital addresses and identifiers that are used in GSM are addressed below.

International Mobile Station Equipment Identity (IMEI)

The International Mobile Station Equipment Identity (IMEI) looks more like a serial number which distinctively identifies a mobile station internationally. This is allocated by the equipment manufacturer and registered by the network operator, who stores it in the Equipment Identity Register (EIR). By means of IMEI, one recognizes obsolete, stolen, or non-functional equipment.

Following are the parts of IMEI:

- Type Approval Code (TAC): 6 decimal places, centrally assigned.

- Final Assembly Code (FAC): 6 decimal places, assigned by the manufacturer.

- Serial Number (SNR): 6 decimal places, assigned by the manufacturer.

- Spare (SP): 1 decimal place.

Thus, IMEI = TAC + FAC + SNR + SP. It uniquely characterizes a mobile station and gives clues about the manufacturer and the date of manufacturing.

International Mobile Subscriber Identity (IMSI)

Every registered user has an original International Mobile Subscriber Identity (IMSI) with a valid IMEI stored in their Subscriber Identity Module (SIM).

IMSI comprises of the following parts:

- Mobile Country Code (MCC): 3 decimal places, internationally standardized.

- Mobile Network Code (MNC): 2 decimal places, for unique identification of mobile network within the country.

- Mobile Subscriber Identification Number (MSIN): Maximum 10 decimal places, identification number of the subscriber in the home mobile network.

Mobile Subscriber ISDN Number (MSISDN)

The authentic telephone number of a mobile station is the Mobile Subscriber ISDN Number (MSISDN). Based on the SIM, a mobile station can have many MSISDNs, as each subscriber is assigned with a separate MSISDN to their SIM respectively.

Listed below is the structure followed by MSISDN categories, as they are defined based on international ISDN number plan:

- Country Code (CC): Up to 3 decimal places.

- National Destination Code (NDC): Typically 2-3 decimal places.

- Subscriber Number (SN): Maximum 10 decimal places.

Mobile Station Roaming Number (MSRN)

Mobile Station Roaming Number (MSRN) is an interim location dependent ISDN number, assigned to a mobile station by a regionally responsible Visitor Location Register (VLA). Using MSRN, the incoming calls are channelled to the MS.

The MSRN has the same structure as the MSISDN.

- Country Code (CC): Of the visited network.

- National Destination Code (NDC): of the visited network.

- Subscriber Number (SN): In the current mobile network.

Location Area Identity (LAI)

Within a PLMN, a Location Area identifies its own authentic Location Area Identity (LAI). The LAI hierarchy is based on international standard and structured in a unique format as mentioned below:

- Country Code (CC): 3 decimal places.

- Mobile Network Code (MNC): 2 decimal places.

- Location Area Code (LAC): maximum 5 decimal places or maximum twice 8 bits coded in hexadecimal (LAC < FFFF).

Temporary Mobile Subscriber Identity (TMSI)

Temporary Mobile Subscriber Identity (TMSI) can be assigned by the VLR, which is responsible for the current location of a subscriber. The TMSI needs to have only local significance in the area handled by the VLR. This is stored on the network side only in the VLR and is not passed to the Home Location Register (HLR).

Together with the current location area, the TMSI identifies a subscriber uniquely. It can contain up to 4×8 bits.

Local Mobile Subscriber Identity (LMSI)

Each mobile station can be assigned with a Local Mobile Subscriber Identity (LMSI), which is an original key, by the VLR. This key can be used as the auxiliary searching key for each mobile station within its region. It can also help accelerate the database access. An LMSI is assigned if the mobile station is registered with the VLR and sent to the HLR. LMSI comprises of four octets (4x8 bits).

Cell Identifier (CI)

Using a Cell Identifier (CI) (maximum 2×8) bits, the individual cells that are within an LA can be recognized. When the Global Cell Identity (LAI + CI) calls are combined, then it is uniquely defined.

GSM – Operations

Once a Mobile Station initiates a call, a series of events takes place. Analyzing these events can give an insight into the operation of the GSM system.

Mobile Phone to Public Switched Telephone Network (PSTN)

When a mobile subscriber makes a call to a PSTN telephone subscriber, the following sequence of events takes place:

- The MSC/VLR receives the message of a call request.

- The MSC/VLR checks if the mobile station is authorized to access the network. If so, the mobile station is activated. If the mobile station is not authorized, then the service will be denied.

- MSC/VLR analyzes the number and initiates a call setup with the PSTN.

- MSC/VLR asks the corresponding BSC to allocate a traffic channel (a radio channel and a time slot).

- The BSC allocates the traffic channel and passes the information to the mobile station.

- The called party answers the call and the conversation takes place.

- The mobile station keeps on taking measurements of the radio channels in the present cell and the neighbouring cells and passes the information to the BSC. The BSC decides if a handover is required. If so, a new traffic channel is allocated to the mobile station and the handover takes place. If handover is not required, the mobile station continues to transmit in the same frequency.

PSTN to Mobile Phone

When a PSTN subscriber calls a mobile station, the following sequence of events takes place:

- The Gateway MSC receives the call and queries the HLR for the information needed to route the call to the serving MSC/VLR.

- The GMSC routes the call to the MSC/VLR.

- The MSC checks the VLR for the location area of the MS.

- The MSC contacts the MS via the BSC through a broadcast message, that is, through a paging request.

- The MS responds to the page request.

- The BSC allocates a traffic channel and sends a message to the MS to tune to the channel. The MS generates a ringing signal and, after the subscriber answers, the speech connection is established.

- Handover, if required, takes place, as discussed in the earlier case.

To transmit the speech over the radio channel in the stipulated time, the MS codes it at the rate of 13 Kbps. The BSC transcodes the speech to 64 Kbps and sends it over a land link or a radio link to the MSC. The MSC then forwards the speech data to the PSTN. In the reverse direction, the speech is received at 64 Kbps at the BSC and the BSC transcodes it to 13 Kbps for radio transmission.

GSM supports 9.6 Kbps data that can be channelled in one TDMA timeslot. To supply higher data rates, many enhancements were done to the GSM standards (GSM Phase 2 and GSM Phase 2+).

GSM - User Services

GSM offers much more than just voice telephony. Contact your local GSM network operator to the specific services that you can avail.

GSM offers three basic types of services:

- Telephony services or teleservices.

- Data services or bearer services.

- Supplementary services.

Teleservices

The abilities of a Bearer Service are used by a Teleservice to transport data. These services are further transited in the following ways:

Voice Calls

The most basic Teleservice supported by GSM is telephony. This includes full-rate speech at 13 kbps and emergency calls, where the nearest emergency-service provider is notified by dialing three digits.

Videotext and Facsmile

Another group of teleservices includes Videotext access, Teletex transmission, Facsmile alternate speech and Facsmile Group 3, Automatic Facsmile Group, 3 etc.

Short Text Messages

Short Messaging Service (SMS) service is a text messaging service that allows sending and receiving text messages on your GSM mobile phone. In addition to simple text messages, other text data including news, sports, financial, language, and location-based data can also be transmitted.

Bearer Services

Data services or Bearer Services are used through a GSM phone. to receive and send data is the essential building block leading to widespread mobile Internet access and mobile data transfer. GSM currently has a data transfer rate of 9.6k. New developments that will push up data transfer rates for GSM users are HSCSD (high speed circuit switched data) and GPRS (general packet radio service) are now available.

Supplementary Services

Supplementary services are additional services that are provided in addition to teleservices and bearer services. These services include caller identification, call forwarding, call waiting, multi-party conversations, and barring of outgoing (international) calls, among others. A brief description of supplementary services is given here:

- Conferencing: It allows a mobile subscriber to establish a multiparty conversation, i.e., a simultaneous conversation between three or more subscribers to setup a conference call. This service is only applicable to normal telephony.

- Call Waiting: This service notifies a mobile subscriber of an incoming call during a conversation. The subscriber can answer, reject, or ignore the incoming call.

- Call Hold: This service allows a subscriber to put an incoming call on hold and resume after a while. The call hold service is applicable to normal telephony.

- Call Forwarding: Call Forwarding is used to divert calls from the original recipient to another number. It is normally set up by the subscriber himself. It can be used by the subscriber to divert calls from the Mobile Station when the subscriber is not available, and so to ensure that calls are not lost.

- Call Barring: Call Barring is useful to restrict certain types of outgoing calls such as ISD or stop incoming calls from undesired numbers. Call barring is a flexible service that enables the subscriber to conditionally bar calls.

- Number Identification: There are following supplementary services related to number identification:

  ◦ Calling Line Identification Presentation: This service displays the telephone number of the calling party on your screen.

  ◦ Calling Line Identification Restriction: A person not wishing their number to be presented to others subscribes to this service.

  ◦ Connected Line Identification Presentation: This service is provided to give the calling party the telephone number of the person to whom they are connected. This service is useful in situations such as forwarding's where the number connected is not the number dialled.

  ◦ Connected Line Identification Restriction: There are times when the person called does not wish to have their number presented and so they would subscribe to this person. Normally, this overrides the presentation service.

  ◦ Malicious Call Identification: The malicious call identification service was provided to combat the spread of obscene or annoying calls. The victim should subscribe to this service, and then they could cause known malicious calls to be identified in the GSM network, using a simple command.

- Advice of Charge (AoC): This service was designed to give the subscriber an indication of the cost of the services as they are used. Furthermore, those service providers who wish to offer rental services to subscribers without their own SIM can also utilize this service in a slightly different form. AoC for data calls is provided on the basis of time measurements.

- Closed User Groups (CUGs): This service is meant for groups of subscribers who wish to call only each other and no one else.

- Unstructured supplementary services data (USSD): This allows operator-defined individual services.

GSM - Security and Encryption

GSM is the most secured cellular telecommunications system available today. GSM has its security methods standardized. GSM maintains end-to-end security by retaining the confidentiality of calls and anonymity of the GSM subscriber.

Temporary identification numbers are assigned to the subscriber's number to maintain the privacy of the user. The privacy of the communication is maintained by applying encryption algorithms and frequency hopping that can be enabled using digital systems and signalling.

This chapter gives an outline of the security measures implemented for GSM subscribers.

Mobile Station Authentication

The GSM network authenticates the identity of the subscriber through the use of a challenge-response mechanism. A 128-bit Random Number (RAND) is sent to the MS. The MS computes the 32-bit Signed Response (SRES) based on the encryption of the RAND with the authentication algorithm (A3) using the individual subscriber authentication key (Ki). Upon receiving the SRES from the subscriber, the GSM network repeats the calculation to verify the identity of the subscriber.

The individual subscriber authentication key (Ki) is never transmitted over the radio channel, as it is present in the subscriber's SIM, as well as the AUC, HLR, and VLR databases. If the received SRES agrees with the calculated value, the MS has been successfully authenticated and may continue. If the values do not match, the connection is terminated and an authentication failure is indicated to the MS.

The calculation of the signed response is processed within the SIM. It provides enhanced security, as confidential subscriber information such as the IMSI or the individual subscriber authentication key (Ki) is never released from the SIM during the authentication process.

Signalling and Data Confidentiality

The SIM contains the ciphering key generating algorithm (A8) that is used to produce the 64-bit ciphering key (Kc). This key is computed by applying the same random number (RAND) used in the authentication process to ciphering key generating algorithm (A8) with the individual subscriber authentication key (Ki).

GSM provides an additional level of security by having a way to change the ciphering key, making the system more resistant to eavesdropping. The ciphering key may be changed at regular intervals as required. As in case of the authentication process, the computation of the ciphering key (Kc) takes place internally within the SIM. Therefore, sensitive information such as the individual subscriber authentication key (Ki) is never revealed by the SIM.

Encrypted voice and data communications between the MS and the network is accomplished by using the ciphering algorithm A5. Encrypted communication is initiated by a ciphering mode request command from the GSM network. Upon receipt of this command, the mobile station begins encryption and decryption of data using the ciphering algorithm (A5) and the ciphering key (Kc).

Subscriber Identity Confidentiality

To ensure subscriber identity confidentiality, the Temporary Mobile Subscriber Identity (TMSI) is used. Once the authentication and encryption procedures are done, the TMSI is sent to the mobile station. After the receipt, the mobile station responds. The TMSI is valid in the location area in which it was issued. For communications outside the location area, the Location Area Identification (LAI) is necessary in addition to the TMSI.

Advantages of GSM

Following are the advantages of GSM:

- GSM techology has been matured since long and hence GSM mobile mobile phones and modems are widely available across the world.

- It provides very cost effective products and solutions.

- The GSM based networks (i.e. base stations) are deployed across the world and hence same mobile phone works across the globe. This leverages cost benefits as well as provides seamless wireless connectivity. This will help users avail data and voice services without any disruption. Hence international roaming is not a concern.

- Advanced versions of GSM with higher number of antennas will provide high speed download and upload of data.

- SAIC and DAIC techniques provide very high transmission quality. SAIC stands for Single Antenna Interference Cancellation technique while DAIC stands for Dual antenna interference cancellation.

- It is easy to maintain GSM networks due to availability of large number of network engineers at affordable cost. This will help in revenue increase by the telecom operators.

- The phone works based on SIM card and hence it is easy to change the different varieties of phones by users.

- The GSM signal does not have any deterioration inside the office and home premises.

- It is easy to integrate GSM with other wireless technology based devices such as CDMA, LTE etc.

Disadvantages of GSM

Following are the disadvantages of GSM:

- Many of the GSM technologies are patented by Qualcomm and hence licenses need to be obtained from them.

- In order to increase the coverage repeaters are required to be installed.

- GSM provides limited data rate capability, for higher data rate GSM advanced version devices are used.

- GSM uses FTDMA access scheme. Here multiple users share same bandwidth and hence will lead to interference when more number of users are using the GSM service. In order to avoid this situation, robust frequency correction algorithms are used in mobile phones and base stations.

- GSM uses pulse based burst transmission technology and hence it interferes with certain electronics. Due to this fact airplanes, petrol bunks and hospitals prevent use of GSM based mobile or other gadgets.

GSM Modem

A GSM modem is a device which can be either a mobile phone or a modem device which can be used to make a computer or any other processor communicate over a network. A GSM modem requires a SIM card to be operated and operates over a network range subscribed by the network operator. It can be connected to a computer through serial, USB or Bluetooth connection.

A GSM modem can also be a standard GSM mobile phone with the appropriate cable and software driver to connect to a serial port or USB port on your computer. GSM modem is usually preferable to a GSM mobile phone. The GSM modem has wide range of applications in transaction terminals, supply chain management, security applications, weather stations and GPRS mode remote data logging.

Working of GSM Modem

From the below circuit, a GSM modem duly interfaced to the MC through the level shifter IC Max232. The SIM card mounted GSM modem upon receiving digit command by SMS from any cell phone send that data to the MC through serial communication. While the program is executed, the GSM modem receives command 'STOP' to develop an output at the MC, the contact point of which are used to disable the ignition switch. The command so sent by the user is based on an intimation received by him through the GSM modem 'ALERT' a programmed message only if the input is driven low. The complete operation is displayed over 16×2 LCD display.

Code Division Multiple Access

Code Division Multiple Access (CDMA) is a digital cellular technology used for mobile communication. CDMA is the base on which access methods such as cdmaOne, CDMA2000, and WCDMA are built. CDMA cellular systems are deemed superior to FDMA and TDMA, which is why CDMA plays a critical role in building efficient, robust, and secure radio communication systems.

Simple Analogy

Let's take a simple analogy to understand the concept of CDMA. Assume we have a few students gathered in a classroom who would like to talk to each other simultaneously. Nothing would be audible if everyone starts speaking at the same time. Either they must take turns to speak or use different languages to communicate.

The second option is quite similar to CDMA — students speaking the same language can understand each other, while other languages are perceived as noise and rejected. Similarly, in radio CDMA, each group of users is given a shared code. Many codes occupy the same channel, but only those users associated with a particular code can communicate.

Salient Features of CDMA

CDMA, which is based on the spread spectrum technique has following salient features:

- In CDMA, every channel uses the full available spectrum.

- Individual conversations are encoded with a pseudo-random digital sequence and then transmitted using a wide frequency range.

- CDMA consistently provides better capacity for voice and data communications, allowing more subscribers to connect at any given time.

- CDMA is the common platform on which 3G technologies are built. For 3G, CDMA uses 1x EV-DO and EV-DV.

Third Generation Standards

CDMA2000 uses Frequency Division Duplexing-Multicarrier (FDD-MC) mode. Here, multicarrier implies N × 1.25 MHz channels overlaid on N existing IS-95 carriers or deployed on unoccupied spectrum. CDMA2000 includes:

- 1x: uses a spreading rate of 1.2288 Mcps.

- 3x: uses a spreading rate of 3 × 1.2288 Mcps or 3.6864 Mcps.

- 1xEV-DO (1x Evolution – Data Optimized): uses a spreading rate of 1.2288 Mcps, optimized for the data.

- WCDMA/FDD-DS: Wideband CDMA (WCDMA) Frequency Division Duplexing-Direct Sequence spreading (FDD-DS) mode. This has a single 5 MHz channel. WCDMA uses a single carrier per channel and employs a spreading rate of 3.84 Mcps.

CDMA Development Group (CDG)

The CDMA Development Group (CDG), founded in December 1993, is an international consortium of companies. It works together to lead the growth and evolution of advanced wireless telecommunication systems.

CDG is comprised of service providers, infrastructure manufacturers, device vendors, test equipment vendors, application developers, and content providers. Its members jointly define the technical requirements for the development of complementary systems CDMA2000 and 4G. Further, the interoperability with other emerging wireless technologies are meant to increase the availability of wireless products and services to consumers and businesses worldwide.

IMT-2000 System

	IMT-DS (Direct Sequence)	IMT-TC m(Multi Carrier)	IMT-TC (Time Code)	IMT-SC (Single Carrier)	IMT-FT (Frequency Time)
Popular name	W-CDMA	CDMA 2000	UTRA-TDD TD-CDMA TD-SCDMA	UWC-136	DECT
Access method	CDMA-FDD	CDMA-FDD	CDMA-TDD	TDMA-FDD	TDMA-TDD
Organization Partners	ARIB/TTC CWTS ESTI T1 TTA	ARIB/TTC CWTS TIA TTA	CWTS ESTI T1 TTA	TIA	EsTI
Body of Technical Spec Production	3GPP(FDD)	3 GPP2	3GPP(TDD) CWTS	IS-136	DECT

CDMA – Channels

CDMA channels can be broadly categorized as Forward channel and Reverse channel. This chapter explains the functionalities of these channels.

Forward Channel

The Forward channel is the direction of the communication or mobile-to-cell downlink path. It includes the following channels:

- Pilot Channel: Pilot channel is a reference channel. It uses the mobile station to acquire the time and as a phase reference for coherent demodulation. It is continuously transmitted by each base station on each active CDMA frequency. And, each mobile station tracks this signal continuously.

- Sync Channel: Synchronization channel carries a single, repeating message, which gives the information about the time and system configuration to the mobile station. Likewise, the mobile station can have the exact system time by the means of synchronizing to the short code.

- Paging Channel: Paging Channel's main objective is to send out pages, that is, notifications of incoming calls, to the mobile stations. The base station uses these pages to transmit system overhead information and mobile station specific messages.

- Forward Traffic Channel: Forward Traffic Channels are code channels. It is used to assign calls, usually voice and signaling traffic to the individual users.

Reverse Channel

The Reverse channel is the mobile-to-cell direction of communication or the uplink path. It consists of the following channels:

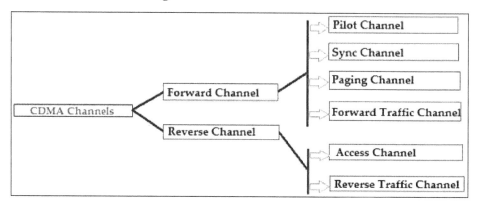

- Access Channel: Access channel is used by mobile stations to establish a communication with the base station or to answer Paging Channel messages. The access channel is used for short signaling message exchanges such as call-ups, responses to pages and registrations.

- Reverse Traffic Channel: Reverse traffic channel is used by the individual users in their actual calls to transmit traffic from a single mobile station to one or more base stations.

FDMA – Technology

Frequency Division Multiple Access (FDMA) is one of the most common analogue multiple access methods. The frequency band is divided into channels of equal bandwidth so that each conversation is carried on a different frequency.

FDMA Overview

In FDMA method, guard bands are used betwent signal spectra to minimize crosstalk between the channels. A specific frequency band is given to one person, and it will received by identifying each of the frequency on the receiving end. It is often used in the first generation of analog mobile phone.

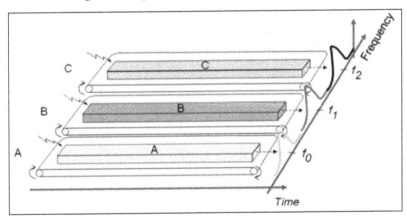

Advantages of FDMA

As FDMA systems use low bit rates (large symbol time) compared to average delay spread, it offers the following advantages:

- Reduces the bit rate information and the use of efficient numerical codes increases the capacity.

- It reduces the cost and lowers the inter symbol interference (ISI).

- Equalization is not necessary.

- An FDMA system can be easily implemented. A system can be configured so that the improvements in terms of speech encoder and bit rate reduction may be easily incorporated.

- Since the transmission is continuous, less number of bits are required for synchronization and framing.

Disadvantages of FDMA

Although FDMA offers several advantages, it has a few drawbacks as well, which are listed below:

- It does not differ significantly from analog systems; improving the capacity depends on the signal-to-interference reduction, or a signal-to-noise ratio (SNR).

- The maximum flow rate per channel is fixed and small.

- Guard bands lead to a waste of capacity.

- Hardware implies narrowband filters, which cannot be realized in VLSI and therefore increases the cost.

TDMA – Technology

Time Division Multiple Access (TDMA) is a digital cellular telephone communication technology. It facilitates many users to share the same frequency without interference. Its technology divides a signal into different timeslots, and increases the data carrying capacity.

TDMA Overview

Time Division Multiple Access (TDMA) is a complex technology, because it requires an accurate synchronization between the transmitter and the receiver. TDMA is used in digital mobile radio systems. The individual mobile stations cyclically assign a frequency for the exclusive use of a time interval.

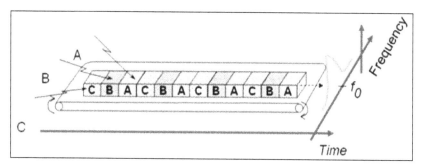

In most of the cases, the entire system bandwidth for an interval of time is not assigned to a station. However, the frequency of the system is divided into sub-bands, and TDMA is used for the multiple access in each sub-band. Sub-bands are known as carrier frequencies. The mobile system that uses this technique is referred as the multi-carrier systems.

In the following example, the frequency band has been shared by three users. Each user is assigned definite timeslots to send and receive data. In this example, user 'B' sends after user 'A,' and user 'C'sends thereafter. In this way, the peak power becomes a problem and larger by the burst communication.

FDMA and TDMA

This is a multi-carrier TDMA system. A 25 MHz frequency range holds 124 single chains (carrier frequencies 200) bandwidth of each kHz; each of these frequency channels contains 8 TDMA conversation channels. Thus, the sequence of timeslots and frequencies assigned to a mobile station is the physical channels of a TDMA system. In each timeslot, the mobile station transmits a data packet.

The period of time assigned to a timeslot for a mobile station also determines the number of TDMA channels on a carrier frequency. The period of timeslots are combined in a so-called TDMA frame. TDMA signal transmitted on a carrier frequency usually requires more bandwidth than FDMA signal. Due to the use of multiple times, the gross data rate should be even higher.

Advantages of TDMA

Here is a list of few notable advantages of TDMA:

- Permits flexible rates (i.e. several slots can be assigned to a user, for example, each time interval translates 32Kbps, a user is assigned two 64 Kbps slots per frame).

- Can withstand gusty or variable bit rate traffic. Number of slots allocated to a user can be changed frame by frame (for example, two slots in the frame 1, three slots in the frame 2, one slot in the frame 3, frame 0 of the notches 4, etc.).

- No guard band required for the wideband system.

- No narrowband filter required for the wideband system.

Disadvantages of TDMA

The disadvantages of TDMA are as follow:

- High data rates of broadband systems require complex equalization.

- Due to the burst mode, a large number of additional bits are required for synchronization and supervision.

- Call time is needed in each slot to accommodate time to inaccuracies (due to clock instability).

- Electronics operating at high bit rates increase energy consumption.

- Complex signal processing is required to synchronize within short slots.

CDMA - Network

CDMA Network is the system meant to regulate CDMA technology. It includes all aspects and functionality starting from the base station, transmitting antenna, receiving antenna, to mobile switching centers.

CDMA Network Overview

A base station is an essential element of the CDMA network. A base station covers a small geographical area called a cell. A cell may be omnidirectional or sectoral. Each base station has a transmitting antenna and two receiving antennas for each cell. Two receiving antennas are used per cell for the purpose of spatial diversity. In many applications, it is a BSC (Base Station Controller), which controls several base stations.

As the rate of the mobile phone data is either 13kbps or 8kbps, which is nonISDN, but the switches which are the mobile switching center (MSC) are generally switched to 64 kbps. Therefore, before it is switched, it is necessary to convert this mobile data rates to 64 kbps. This is accomplished by a member, which is the transcoder. The transcoder may be a separate element or it can be collocated in each base station or MSC.

All base stations are connected to the MSC, which is the mobile switching center. MSC is the entity that manages the establishment, connection, maintenance, and disposal of calls within the network and also with the outside world.

MSC also has a database called HLR/AC, which is a home location register/authentication center. HLR is the database, which maintains the database of all network subscribers. AC Authentication Centre is the part of the security of the HLR, which some algorithms to examine mobile phones.

The MSC is connected to the outside world, i.e. the fixed line network. MSC can also be connected to several other MSCs.

CDMA Identities

Network Identities:

- SID (System Identity).

- NID (Network Identity).

Mobile Station Identities:

- ESN (Electronic Serial Number).

- Permuted ESN.

- IMSI (International Mobile Station Identity).

- IMSI_S.

- IMSI_11_12.

- Station Class Mark.

System and Network Identity

A base station is a member of a cellular system and a network. A network is a subset of a system. The systems are installed with an identification called Identification System (CIS). The networks with a system receiving is Network identification (NID). It is a uniquely identified network pair of (SID, NID). The mobile station has a list of one or more home (non-roaming) pairs (SID, NID).

SID

A system identification indicator 15 bits (SID) is stored in a mobile station. It is used to determine the host system of the mobile stations. The bit allocation of the system identification indicator is shown below.

14 Bit	13 Bit	12 Bit	0 Bit
INTL CODE		SYSTEM NUMBER	

The distribution of international codes (INTL) (bits 14 and 13) is also shown in the table. Bits 12-0 is assigned to each US system by the FCC for non-US countries. The bit allocation will be made by local regulatory authorities.

NID

NID has a range of 0-65535 reserved values. Value of 65535 in a SID means, NID pair is to indicate that the Mobile Station considers the entire SID as home.

65535	0
SYSTEM NUMBER	

Systems and Networks

A mobile station has a list of one or more home (non-roaming) pairs (SID, NID). A mobile station is roaming when the base station broadcast (SID, NID) pair does not match with one of the non-roaming mobile stations (SID, NID) pairs.

A mobile station is a foreign NID roamer:

- If the mobile station is roaming and there are some (SID, NID) pair in the mobile stations (SID, NID) list that corresponds to SID.

- If the mobile station is roaming and there are some (SID, NID) pair in the mobile stations (SID, NID) list for which no matching SID is available (means a mobile station has roaming customer foreign SID).

Electronic Serial Number (ESN)

ESN is a 32-bit binary number that uniquely identifies the mobile station in a CDMA cellular system. It should be set at the factory and cannot be easily changed in the field. Changing the ESN will require special equipment, not normally available to subscribers. The bit allocation of ESN is shown below:

31 Bit	24 Bit	23 Bit	17 Bit	18 Bit	0 Bit
MFR CODE		RESERVED VALUE		SERIAL NUMBER	

The circuit that provides the ESN must be isolated so that no one can contact and tamper. Attempts to change the ESN circuit should make the mobile station inoperative. At the time of the issuance of the initial acceptance, the manufacturer must be assigned a code Manufacturers (MFR) in the eight most significant bits (bits 31-24 bits) 32-bit serial number. Bits 23-18 are reserved (initially zero). And, every manufacturer only allocates 17 bits to 0. When a manufacturer has used almost all possible combinations of serial numbers in bits 17-0, the manufacturer may submit a notification to the FCC. The FCC will assign the next sequential binary number in the reserve block (bits 23 through).

Permuted ESN

CDMA is a spread spectrum technique where multiple users to access the system at the same example in a cell, and of course on the same frequency. Therefore, it discriminates the users on the reverse link (i.e. information from MS to the base station). It spreads information using codes that are unique to the mobile station in all the CDMA cellular systems. This code has an element that is the ESN, but it doesn't use the ESN in the same format instead, it uses an ESN swapped.

If there are two mobiles in a cell of the same brand and have consecutive serial numbers and for the receiver of the base station, it becomes difficult to connect them. Therefore, to avoid a strong correlation between the long codes corresponding to successive ESN, we use permuted ESNs.

International Mobile Station Identity (IMSI)

Mobile stations are identified by the identity of the international mobile station Identity (IMSI). The IMSI consists of up to 10 to 15 numeric digits. The first three digits of the IMSI are the country code of the mobile (MCC), the remaining digits are the National NMSI mobile station identity. The NMSI consists of the mobile network code (MNC) and the mobile station identification number (SIDS).

MCC	MSN	MSIN
	NMSI	
IMSI ≤15 digits		

- MCC: Mobile Country Code.

- MNC: Mobile Network Code.

- MSIN: Mobile Station Identification.

- NMSI: National Mobile Station Identity.

An IMSI that is 15 digits in length is called a class 0 IMSI (NMSI is the 12 digits in length). IMSI, which is less than 15 digits in length, is called a class 1 IMSI (NMSI the length is less than 12 counts). For CDMA operation, the same IMSI may be registered in multiple mobile stations. Individual systems may or may not allow these capabilities. The management of these functions is a function of the base station and the system operator.

General Packet Radio Service

GPRS is an expansion Global System for Mobile Communication. It is basically a packet-oriented mobile data standard on the 2G and 3G cellular communication network's global system for mobile communication. GPRS was built up by European Telecommunications Standards Institute (ETSI) because of the prior CDPD, and I-mode packet switched cell advances.

GPRS overrides the wired associations, as this framework has streamlined access to the packet information's network like the web. The packet radio standard is utilized by GPRS to transport client information packets in a structured route between GSM versatile stations and external packet information networks. These packets can be straightforwardly directed to the packet changed systems from the GPRS portable stations.

GPRS is also known as GSM-IP that is a Global-System Mobile Communications Internet Protocol as it keeps the users of this system online, allows to make voice calls, and access internet on-the-go. Even Time-Division Multiple Access (TDMA) users benefit from this system as it provides packet radio access.

GPRS also permits the network operators to execute an Internet Protocol (IP) based core architecture for integrated voice and data applications that will continue to be used and expanded for 3G services.

GPRS supersedes the wired connections, as this system has simplified access to the packet data networks like the internet. The packet radio principle is employed by GPRS to transport user data packets in a structure way between GSM mobile stations and external packet data networks. These packets can be directly routed to the packet switched networks from the GPRS mobile stations.

In the current versions of GPRS, networks based on the Internet Protocol (IP) like the global internet or private/corporate intranets and X.25 networks are supported.

Key Features

Following three key features describe wireless packet data:

- The always online feature: Removes the dial-up process, making applications only one click away.

- An upgrade to existing systems: Operators do not have to replace their equipment; rather, GPRS is added on top of the existing infrastructure.

- An integral part of future 3G systems: GPRS is the packet data core network for 3G systems EDGE and WCDMA.

Goals of GPRS

GPRS is the first step toward an end-to-end wireless infrastructure and has the following goals:

- Open architecture.

- Consistent IP services.

- Same infrastructure for different air interfaces.

- Integrated telephony and Internet infrastructure.

- Leverage industry investment in IP.

- Service innovation independent of infrastructure.

Benefits of GPRS

Higher Data Rate

GPRS benefits the users in many ways, one of which is higher data rates in turn of shorter access times. In the typical GSM mobile, setup alone is a lengthy process and equally, rates for data permission are restrained to 9.6 kbit/s. The session establishment time offered while GPRS is in practice is lower than one second and ISDN-line data rates are up to many 10 kbit/s.

Easy Billing

GPRS packet transmission offers a more user-friendly billing than that offered by circuit switched services. In circuit switched services, billing is based on the duration of the connection. This is unsuitable for applications with bursty traffic. The user must pay for the entire airtime, even for idle periods when no packets are sent (e.g., when the user reads a Web page).

In contrast to this, with packet switched services, billing can be based on the amount of transmitted data. The advantage for the user is that he or she can be "online" over a long period of time but will be billed based on the transmitted data volume.

GPRS - Architecture

GPRS architecture works on the same procedure like GSM network, but, has additional entities that allow packet data transmission. This data network overlaps a second-generation GSM network providing packet data transport at the rates from 9.6 to 171 kbps. Along with the packet data transport the GSM network accommodates multiple users to share the same air interface resources concurrently.

Following is the GPRS Architecture diagram:

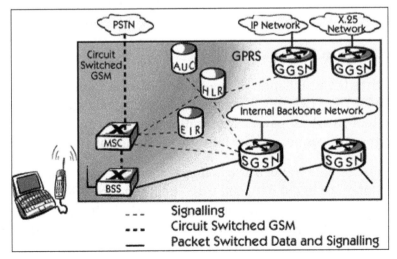

GPRS attempts to reuse the existing GSM network elements as much as possible, but to effectively build a packet-based mobile cellular network, some new network elements, interfaces, and protocols for handling packet traffic are required.

Therefore, GPRS requires modifications to numerous GSM network elements as summarized below:

GSM Network Element	Modification or Upgrade Required for GPRS.
Mobile Station (MS)	New Mobile Station is required to access GPRS services. These new terminals will be backward compatible with GSM for voice calls.
BTS	A software upgrade is required in the existing Base Transceiver Station(BTS).
BSC	The Base Station Controller (BSC) requires a software upgrade and the installation of new hardware called the packet control unit (PCU). The PCU directs the data traffic to the GPRS network and can be a separate hardware element associated with the BSC.
GPRS Support Nodes (GSNs)	The deployment of GPRS requires the installation of new core network elements called the serving GPRS support node (SGSN) and gateway GPRS support node (GGSN).
Databases (HLR, VLR, etc.)	All the databases involved in the network will require software upgrades to handle the new call models and functions introduced by GPRS.

GPRS Mobile Stations

New Mobile Stations (MS) are required to use GPRS services because existing GSM phones do not handle the enhanced air interface or packet data. A variety of MS can exist, including a high-speed version of current phones to support high-speed data access, a new PDA device with an embedded GSM phone, and PC cards for laptop computers. These mobile stations are backward compatible for making voice calls using GSM.

GPRS Base Station Subsystem

Each BSC requires the installation of one or more Packet Control Units (PCUs) and a software upgrade. The PCU provides a physical and logical data interface to the Base Station Subsystem (BSS) for packet data traffic. The BTS can also require a software upgrade but typically does not require hardware enhancements.

When either voice or data traffic is originated at the subscriber mobile, it is transported over the air interface to the BTS, and from the BTS to the BSC in the same way as a standard GSM call. However, at the output of the BSC, the traffic is separated; voice is sent to the Mobile Switching Center (MSC) per standard GSM, and data is sent to a new device called the SGSN via the PCU over a Frame Relay interface.

GPRS Support Nodes

Following two new components, called Gateway GPRS Support Nodes (GSNs) and, Serving GPRS Support Node (SGSN) are added:

Gateway GPRS Support Node (GGSN)

The Gateway GPRS Support Node acts as an interface and a router to external networks. It contains routing information for GPRS mobiles, which is used to tunnel packets through the IP based internal backbone to the correct Serving GPRS Support Node. The GGSN also collects charging information connected to the use of the external data networks and can act as a packet filter for incoming traffic.

Serving GPRS Support Node (SGSN)

The Serving GPRS Support Node is responsible for authentication of GPRS mobiles, registration of mobiles in the network, mobility management, and collecting information on charging for the use of the air interface.

Internal Backbone

The internal backbone is an IP based network used to carry packets between different GSNs. Tunnelling is used between SGSNs and GGSNs, so the internal backbone does not need any information about domains outside the GPRS network. Signalling from a GSN to a MSC, HLR or EIR is done using SS7.

Routing Area

GPRS introduces the concept of a Routing Area. This concept is similar to Location Area in GSM, except that it generally contains fewer cells. Because routing areas are smaller than location areas, less radio resources are used While broadcasting a page message.

GPRS – Applications

GPRS has opened a wide range of unique services to the mobile wireless subscriber. Some of the characteristics that have opened a market full of enhanced value services to the users. Below are some of the characteristics:

- Mobility - The ability to maintain constant voice and data communications while on the move.

- Immediacy - Allows subscribers to obtain connectivity when needed, regardless of location and without a lengthy login session.

- Localization - Allows subscribers to obtain information relevant to their current location.

Using the above three characteristics varied possible applications are being developed to offer to the mobile subscribers. These applications, in general, can be divided into two high-level categories:

- Corporation.

- Consumer.

These two levels further include:

- Communications: E-mail, fax, unified messaging and intranet/internet access, etc.

- Value-added services: Information services and games, etc.

- E-commerce: Retail, ticket purchasing, banking and financial trading, etc.

- Location-based applications: Navigation, traffic conditions, airline/rail schedules and location finder, etc.

- Vertical applications: Freight delivery, fleet management and sales-force automation.

- Advertising: Advertising may be location sensitive. For example, a user entering a mall can receive advertisements specific to the stores in that mall.

Along with the above applications, non-voice services like SMS, MMS and voice calls are also possible with GPRS. Closed User Group (CUG) is a common term used after GPRS is in the market, in addition, it is planned to implement supplementary services, such as Call Forwarding Unconditional (CFU), and Call Forwarding on Mobile subscriber Not Reachable (CFNRc), and closed user group (CUG).

Universal Mobile Telecommunications Service

The Universal Mobile Telecommunications System (UMTS) is a third generation mobile cellular system for networks based on the GSM standard. Developed and maintained by the 3GPP (3rd Generation Partnership Project), UMTS is a component of the Standard International Union all IMT-2000 telecommunications and compares it with the standard set for CDMA2000 networks based on competition cdmaOne technology. UMTS uses wideband code division multiple access (W-CDMA) radio access technology to provide greater spectral efficiency and bandwidth mobile network operators.

Network Evolution

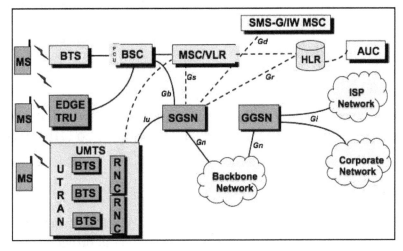

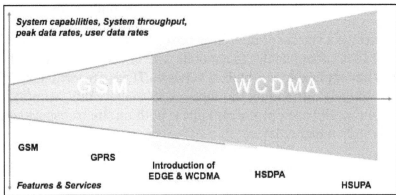

HSUPA – High Speed Uplink Packet Access

HSDPA – High Speed Downlink Packet Access

The main idea behind 3G is to prepare a universal infrastructure able to carry existing and also future services. The infrastructure should be so designed that technology changes and evolution can be adapted to the network without causing uncertainties to the existing services using the existing network structure.

UMTS – Objectives

UMTS – Radio Interface and Radio Network Aspects

After the introduction of UMTS the amount of wide area data transmission by mobile users had picked up. But for the local wireless transmissions such as WLAN and DSL, technology has increased at a much higher rate. Hence, it was important to consider the data transmission rates equal to the category of fixed line broadband, when WIMAX has already set high targets for transmission rates. It was clear that the new 3GPP radio technology Evolved UTRA (E-UTRA, synonymous with the LTE radio interface) had to become strongly competitive in all respect and for that following target transmission rates were defined:

- Downlink: 100 Mb/s.

- Uplink: 50 Mb/s.

Above numbers are only valid for a reference configuration of two antennas for reception and one transmit antenna in the terminal, and within a 20 MHz spectrum allocation.

UMTS – All IP Vision

A very general principle was set forth for the Evolved 3GPP system. It should "all IP", means that the IP connectivity is the basic service which is provided to the users. All other layer services like voice, video, messaging, etc. are built on that.

Looking at the protocol stacks for interfaces between the network nodes, it is clear that simple model of IP is not applicable to a mobile network. There are virtual layers in between, which is not applicable to a mobile network. There are virtual layer in between, in the form of "tunnels", providing the three aspects - mobility, security, and quality of service. Resulting, IP based protocols appear both on the transport layer (between network nodes) and on higher layers.

UMTS – Requirements of the New Architecture

There is a new architecture that covers good scalability, separately for user plane and control plane. There is a need for different types of terminal mobility support that are: fixed, nomadic, and mobile terminals.

The minimum transmission and signaling overhead especially in air, in an idle mode of the dual mode UE signaling should be minimized, in the radio channel multicast capability. It is required to be reused or extended, as roaming and network sharing restrictions, compatible with traditional principles established roaming concept, quite naturally, the maximum transmission delay required is equivalent to the fixed network, specifically less than 5 milliseconds, set to control plane is less than 200 milliseconds delay target.

Looking at the evolution of the 3GPP system in full, it may not seem less complex than traditional 3GPP system, but this is due to the huge increase in functionality. Another strong desire is to arrive at a flat structure, reducing CAPEX/OPEX for operators in the 3GPP architecture carriers.

Powerful control functions should also be maintained with the new 3GPP systems, both real-time seamless operation (for example, VoIP) and non-real-time applications and services. The system should perform well for VoIP services in both the scenarios. Special attention is also paid to the seamless continuity with legacy systems (3GPP and 3GPP2), supports the visited network traffic local breakout of voice communications.

UMTS – Security and Privacy

Visitor Location Register (VLR) and SNB are used to keep track of all the mobile stations that are currently connected to the network. Each subscriber can be identified by its International Mobile Subscriber Identity (IMSI). To protect against profiling attacks, the permanent identifier is sent over the air interface as infrequently as possible. Instead, local identities Temporary Mobile Subscriber force (TMSI) is used to identify a subscriber whenever possible. Each UMTS subscriber has a dedicated home network with which it shares a secret key Ki long term.

The Home Location Register (HLR) keeps track of the current location of all the home network subscribers. Mutual authentication between a mobile station and a visited network is carried out with the support of the current GSN (SGSN) and the MSC / VLR, respectively. UMTS supports encryption of the radio interface and the integrity protection of signaling messages.

UMTS – Authentication

UMTS is designed to interoperate with GSM networks. To protect GSM networks against man-in-middle attacks, 3GPP is considering to add a structure RAND authentication challenge.

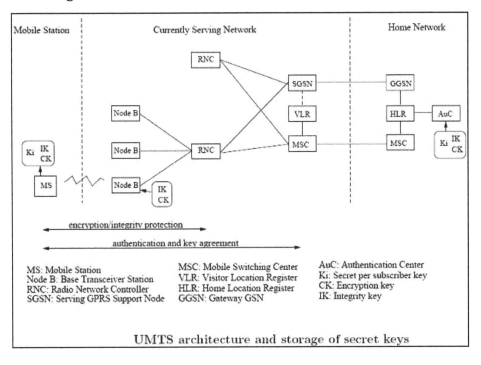

UMTS architecture and storage of secret keys

UMTS Subscriber to UMTS Network

Both the network and the mobile station supports all the security mechanisms of UMTS. Authentication and Key agreement is as follows:

- The mobile station and the base station to establish a radio resource control connection (RRC connection). During the establishment of the connection the mobile station sends its security capabilities to the base station. Security features include UMTS integrity and encryption algorithms supported and possibly GSM encryption capabilities as well.

- The mobile station sends its temporary identity TMSI current on the network.

- If the network cannot solve the TMSI, he asks the mobile station to send its permanent identity and the mobile stations responding to the request with the IMSI.

- The visited network requests authentication of the home network of the mobile station data.

- The home network returns a random challenge RAND, the corresponding authentication token AUTN, authentication

- Response XRES, integrity key IK and the encryption key CK.

- The visited network sends RAND authentication challenge and authentication token AUTN to the mobile Station.

- The mobile station checks AUTN and calculates the authentication response. If AUTN is corrected.

- Mobile station ignores the message.

- The mobile station sends its authentication response RES to the visited network.

- Visiting the network checks if RES = XRES and decide which security algorithms radio subsystem is allowed to use.

- The visited network sends algorithms admitted to the radio subsystem.

- The radio access network decides permit (s) algorithms to use.

- The radio access network informs the mobile station of their choice in the security mode command message.

- The message also includes the network security features received from the mobile station in step 1.

- This message is integrity protected with the integrity key IK.

- The mobile station confirms the protection of the integrity and verify the accuracy of the safety functions.

UMTS Subscriber to GSM Base Station

The mobile unit (subscriber UMTS) supports both USIM and SIM application. The base station system uses GSM while the VLR / MSC technology components are respectively the UMTS SGSN. The mobile station and the core network both support all security mechanisms of UMTS. However, the base station system GSM (BSS) does not support the protection of the integrity and uses the GSM encryption algorithms. The first eight steps of the authentication protocol are performed as in the classical case. GSM BSS simply forwards the UMTS authentication traffic.

- The MSC / SGSN decides which GSM encryption algorithms are allowed and calculates the key GSM Kc UMTS keys IK, CK.

- The MSC / SGSN advises the GSM BSS authorized algorithms and transmits the GSM cipher key Kc.

- GSM BSS decide which encryption algorithms allowed to use based encryption capabilities of the mobile station.

- GSM BSS sends the GSM cipher mode command to the station.

Next Generation Network

Next generation network (NGN) is a broad term used to describe architectural evolution and innovation in telecommunication and access networking. Next generation networks are used to transmit all kinds of services and information including voice data/calls, audio data/calls and multimedia information such as videos. All kinds of such data are encapsulated in data packet form.

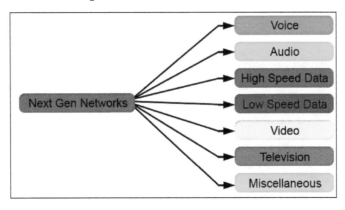

NGN is a packet-based network that provides services like broadband, telecommunication and is also able to use QoS.

A NGN is a purely based on Internet technology, including MPLS and IP. It uses H.323 protocol as its major component. Softswitch is the most important device used in NGN and it is only designed for voice applications.

IP Multimedia Subsystem (IMS) is a global IP based service architecture offering various multimedia services. It is standardized by 3GPP (3rd Generation Partnership Project) and referred to be the heart of NGN.

Technology

A NGN converges the Service Provider networks including the Public Switched Telephone Network (PSTN), data network (the Internet), and wireless network. It offers a high quality end user experience. But the most critical challenge is in optimizing the OSS and BSS platforms, systems, and processes at various levels such as the Fixed Line incumbents, Mobile operators, Cable TV operators, Unified Access Service Providers, Internet Service Providers, Software and Hardware vendors etc.

At core network, NGN consolidates several transport networks into one core transport network based on IP and Ethernet with migrations from PSTN to VoIP, legacy services of X.25 and Frame Relay to IP VPN. At wired access network, NGN is responsible for the migration from dual legacy voice next to xDSL setup to a converged setup. At cable access network, NGN convergence involves migrating from bit-rate voice to standards like VoIP and SIP. MGN architecture as defined in ITU-T Rec. Y.2012 is given below.

Functional architecture of NGN is given below. It shows four different layers.

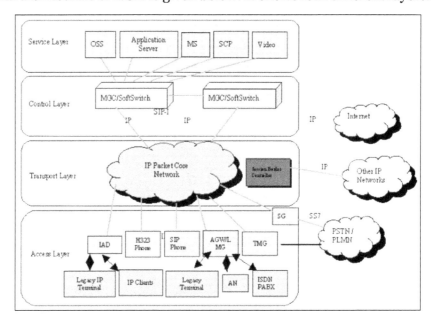

- Access Layer: Connects subscribers PSTN, ISDN, PLMN etc and converts information formats (circuit-to-packet, packet-to-circuit).

- Transport Layer: Offers connectivity for all components and supports transfer of information.

- Control Layer: Offers software switching to achieve real time call control, media gateway access control, resource allocation, protocol processing, routing, and authentication.

- Service Layer: Offers many value-added services such as supporting operating system, managing application, video, and media servers.

Advantages

NGN offers several advantages at various levels of services:

- In Unified Messaging, it supports the transmission of voice mail, email, fax mail, and pages through common interfaces.

- In Data Connectivity, it offers many value added services such as bandwidth on demand, durable Switched Virtual Connections (SVC), call admission control etc.

- In Voice Telephony, it supports all traditional telephony services besides focusing on the most marketable voice telephony features.

- In Multimedia, it enables collaborative computing and groupware and supports interactivity among multiple parties sharing voice, video, and/or data.

- In Public Network Computing (PNC), it supports generic processing and storage capabilities, Enterprise Resource Planning (ERP), time reporting, and miscellaneous consumer applications.

- In Home Networking, it supports intelligent appliances, home security systems, energy systems, and entertainment systems.

- In Virtual Call Centers, it enables voice calls and e-mail messages through queue system, electronic access to customer, catalog, stock, and ordering information, and communication between customer and agent.

- In Information Brokering, it offers advertising and information delivery based on pre-specified criteria or personal preferences and behavior patterns.

- In Interactive Gaming, it establishes interactive gaming sessions among multiple users.

- In Virtual Private Network (VPN), it offers uniform dialing capabilities for voice VPNs and added security and network features for data VPNs.

- In Ecommerce, it enables e-transactions, verification of payment information, trading, home banking and shopping etc.

- In Distributed Virtual Reality, it builds up co-ordination among multiple diverse resources in providing real world events, people, places, experiences, etc.

BcN Model

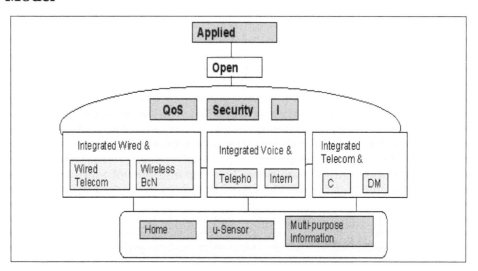

The first step of BcN implementation was planned to be completed by 2005. The main tasks of this step are:

- Integrating Voice and Data network based on Open Architecture.

- Establishing service convergence between Wired and Wireless.

- Expanding FTTC (VDSL/HFC).

- Introducing new services like FTTH, Terrestrial DMB, Satellite DMB, and IPv6 based Home Network.

The second step of BcN implementation was planned to be completed by 2007. The main tasks of this step are:

- Converging of Wired & Wireless network.

- Implementing Transport layer based IPv6/MPLS.

- Establishing Service Convergence between Communication & Broadcasting.

- Expanding FTTH.

- Introducing New Services like HPi and interactive DMB, Ubiquitous Sensor network.

The third step of BcN implementation was planned to be completed by 2010. The main tasks of this step are:

Converging Services like voice network, internet, mobile network, broadcasting, high speed data network etc., over a single Transport Layer. Supporting New Service requirements such as broadband, QoS, security, mobility, and multicasting.

Some of BCN's trial services are given below:

- Trial service from Octave Consortium of Korean Telecom (KT) covers 600 subscribers across 3 cities of Seoul, Daejeon, and Daegu with 25 different services of Telephony data and applications.

- Trial service from UbiNet Consortium of SKT, Hanaro Telecom covers 600 subscribers across 3 cities of Seoul, Busan, and Daegu with 32 different services.

- Trial service from Gwanggaeto Consortium of Dacom covers 350 subscribers in 5 areas of Seoul, Gyunggi, Bucheon, Busan, and Gwangju with 25 different services.

- Trial service from Cable BcN Consortium of Cable Providers covers 700 subscribers in 7 areas.

References

- Mobile-communication-protocols, communication-technologies: tutorialspoint.com, Retrieved 02 June, 2019

- Gsm-architecture-features-working: elprocus.com, Retrieved 16 June, 2019

- Network-architecture, connectivity-2g-gsm: electronics-notes.com, Retrieved 14 April, 2019

- Gsm-specification, gsm: tutorialspoint.com, Retrieved 03 February, 2019

- Advantages-and-Disadvantages-of-GSM: rfwireless-world.com, Retrieved 20 January, 2019

- Gsm-architecture-features-working: elprocus.com, Retrieved 25 July, 2019

- General-packet-radio-service-gprs: geeksforgeeks.org, Retrieved 16 March, 2019

- Next-generation-network-ngn- 25065: techopedia.com, Retrieved 07 August, 2019

- Next-generation-networking: ipv6.com, Retrieved 09 June, 2019

Permissions

All chapters in this book are published with permission under the Creative Commons Attribution Share Alike License or equivalent. Every chapter published in this book has been scrutinized by our experts. Their significance has been extensively debated. The topics covered herein carry significant information for a comprehensive understanding. They may even be implemented as practical applications or may be referred to as a beginning point for further studies.

We would like to thank the editorial team for lending their expertise to make the book truly unique. They have played a crucial role in the development of this book. Without their invaluable contributions this book wouldn't have been possible. They have made vital efforts to compile up to date information on the varied aspects of this subject to make this book a valuable addition to the collection of many professionals and students.

This book was conceptualized with the vision of imparting up-to-date and integrated information in this field. To ensure the same, a matchless editorial board was set up. Every individual on the board went through rigorous rounds of assessment to prove their worth. After which they invested a large part of their time researching and compiling the most relevant data for our readers.

The editorial board has been involved in producing this book since its inception. They have spent rigorous hours researching and exploring the diverse topics which have resulted in the successful publishing of this book. They have passed on their knowledge of decades through this book. To expedite this challenging task, the publisher supported the team at every step. A small team of assistant editors was also appointed to further simplify the editing procedure and attain best results for the readers.

Apart from the editorial board, the designing team has also invested a significant amount of their time in understanding the subject and creating the most relevant covers. They scrutinized every image to scout for the most suitable representation of the subject and create an appropriate cover for the book.

The publishing team has been an ardent support to the editorial, designing and production team. Their endless efforts to recruit the best for this project, has resulted in the accomplishment of this book. They are a veteran in the field of academics and their pool of knowledge is as vast as their experience in printing. Their expertise and guidance has proved useful at every step. Their uncompromising quality standards have made this book an exceptional effort. Their encouragement from time to time has been an inspiration for everyone.

The publisher and the editorial board hope that this book will prove to be a valuable piece of knowledge for students, practitioners and scholars across the globe.

Index